OECD Studies on Water

Mitigating Droughts and Floods in Agriculture

POLICY LESSONS AND APPROACHES

Please cite this publication as:
OECD (2016), *Mitigating Droughts and Floods in Agriculture: Policy Lessons and Approaches*, OECD Studies on Water, OECD Publishing, Paris.
http://dx.doi.org/10.1787/9789264246744-en

ISBN 978-92-64-24673-7 (print)
ISBN 978-92-64-24674-4 (PDF)

Series: OECD Studies on Water
ISSN 2224-5073 (print)
ISSN 2224-5081 (online)

Revised version, May 2017
Details of revisions available at: *www.oecd.org/about/publishing/Corrigendum-Mitigating-Droughts-Floods-Agriculture.pdf*

Photo credits: Cover © iStockphoto/sbayram

Corrigenda to OECD publications may be found on line at: *www.oecd.org/about/publishing/corrigenda.htm*.

Foreword

Agriculture will face multiple challenges in the 21st century to provide food, fibre and energy for a growing population whose eating habits are rapidly changing. It must do so while improving its environmental performance and sustainability in a context of climate change.

Climatic conditions have always mattered for the agricultural sector. Farmers have dealt with weather hazards with remarkable ingenuity, and policy makers have developed responses to manage such risks, and indeed to anticipate and manage them more effectively. This, in turn, has greatly contributed to the considerable progress made in terms of agricultural productivity and water management. However, more must still be done and particularly in light of the growing constraints and unknowns imposed by climate change.

This OECD report proposes a comprehensive analysis of, and a set of key recommendations on policy approaches to the sustainable management of droughts and floods in agriculture. It builds on recent trends, experiences and research from OECD countries in this area, in particular Australia, Canada, France, Spain and the United Kingdom. Beyond policy recommendations, this report also provides a general policy framework that could be useful for countries to analyse their own drought and flood policies, as well as to identify ways forward. This report is part of a wider set of OECD work on risk management in agriculture, which can be found on the dedicated webpage: www.oecd.org/tad/agricultural-policies/risk-management-agriculture.htm.

Acknowledgements

This report was drafted by Julien Hardelin and Jesús Antón, of the OECD. It builds on useful advice, recommendations, and contributions provided by the following consultants: Professor Alberto Garrido and Dolores Rey from the Research Centre for the Management of Agricultural and Environmental Risks (CEIGRAM), Technical University of Madrid; Stefan Ambec, INRA Research Professor and Arnaud Reynaud, INRA Researcher, Director at the Laboratoire d'Économie des Ressources Naturelles (LERNA), Toulouse School of Economics (TSE); Christine Le Thi, of the OECD, also contributed to the project in its early stages. Many OECD colleagues contributed to the report, in particular, Charles Baubion, Françoise Bénicourt, Carmel Cahill, Anthony Cox, Kathleen Dominique, Marina Giacalone-Belkadi, Guillaume Gruère, Franck Jésus, Hannah Leckie, Stéphanie Lincourt, Xavier Leflaive, Kevin Parris, Michèle Patterson, Janine Treves, and Frank Van Tongeren. This report was declassified by the OECD Joint Working Party on Agriculture and the Environment in August 2015.

Table of contents

Tables

Figures

Boxes

Executive summary

The agriculture sector is particularly exposed to risks of floods and droughts, which may become more frequent and severe due to climate change in a context of increased demand for food and urban space. In the course of the 20th century, agricultural productivity growth and policy development have allowed to better cope with these risks and reduce their overall impact on the economy and the agricultural sector itself. They nevertheless remain an important issue in many countries, causing substantial costs to the sector and, in some cases, impacts on commodity markets. Three major drivers will likely make drought and flood risks a growing policy concern in the future:

- Increased population and associated rising demand for food, feed, fibre, and energy in the context of rising competition for water resources and increasing vulnerability to water shortages and drought risks.
- Urbanisation will increase the demand for flood protection and mitigation, raising the issue of the allocation of flood risks across sectors and areas, including agricultural lands.
- Climate change is expected to increase the frequency and magnitude of extreme weather events.

Governments play a key role in developing targeted policy responses to market failures that impede the efficient mitigation and allocation of drought and flood risks. However, OECD country experiences suggest there is substantial room to improve policy responses. Indeed, public policies in the domains of water rights and water allocation, weather and hydrological information, innovation and education, and insurance and compensation against drought and flood risks should be combined in a more consistent way to foster efficient management of these risks.

When possible, governments should undertake cost-benefit analysis of policy responses that address droughts and floods risks at the national or even regional levels, and use state-of-the-art methodologies to ensure that local specificities are accounted for and thus avoid possible unintended side-effects. Indeed, the physical characteristics of floods and droughts (frequency, magnitude, and perimeter), and the exposure and vulnerability of the agriculture sector to these risks vary a great deal across and within countries. Hence, it is important to understand the local context when considering policy responses as they will need to be tailored to specific circumstances.

Agricultural policy incentives should avoid supporting production decisions that increase exposure and vulnerability of agricultural systems to droughts and floods. Distortionary subsidies, such as guaranteed prices or insurance subsidies, may lead farmers to artificially increase their risk exposure, and thus their risk vulnerability in the long term. Reducing such distorting subsidies will ensure that production and investment decisions by farmers are based on the real costs and benefits of risk taking.

Ensuring water rights that reflect water availability within sustainable limits is a prerequisite to any coherent policy to managing droughts in agriculture. Over-allocation of water rights and lack of incentives to reflect the scarcity cost of water leads to structural water deficits and chronic water shortages, mechanically increasing exposure and vulnerability to drought risks.

To balance aggregate water demand and supply, water use efficiency and hydrological infrastructures are key elements of a drought mitigation strategy, especially in countries with arid and semi-arid climates. The economic, social and environmental performance of hydrological infrastructure (dams, water reservoirs and alternative sources of water supply) has been often questioned. In the context of climate change, hydrological infrastructures could be further developed in some regions to increase water supply and mitigate exposure and vulnerability of water users as part of a water management strategy. However, policies fostering water use efficiency and water storage need to be compatible with economic, social and environmental objectives and with sustainable production growth. In particular, development of hydrological infrastructure, if desirable in certain cases, should complement rather than replace water policies on the demand side. Improvement of irrigation efficiency and the development of water-saving farming practices can be more efficient to balance aggregate water supply and demand. Ultimately, any cost-benefit analysis of hydrological infrastructure should take into account the deep uncertainty about the projected impacts of climate change at the local scale and the associated option values.

Properly designed water allocation systems can also mitigate the overall cost of droughts to irrigated agriculture by ensuring that water withdrawals are directed to the most socially valuable uses. This would include the use of economic instruments and weather and hydrological information systems. Economic instruments such as water markets have the potential to mitigate substantially the costs of water shortages as demonstrated in several OECD countries that implemented such tools. Weather and hydrological information systems, already well developed in most OECD countries, are essential.

As regards flood management, OECD countries could explore more systematically the potential benefits of agricultural land areas as a provider of flood control services. Such areas can be used as floodplains and providers of water retentions services, thus mitigating flood risks in other areas such as cities and consequently reducing the overall cost of flood risks. Moreover, floodplains could also be sources of additional ecosystem services that should be included in cost-benefit analysis. However, this should be a complementary tool rather than a substitute to proper urban planning processes.

Short-term water risk management is also an important dimension of an efficient policy approach to managing droughts and floods in agriculture:

- Weather and hydrological information systems are crucial to ensure farmers have the possibility adapt their production plan at the beginning of the cropping season.
- As regards droughts, developing flexible water reallocation rules based on aggregate water availability, including a combination of priority rules and short-term water markets in agriculture. Such tools can help adapt water demand to changing weather and hydrological conditions and integrate both marketable and non-marketable water uses (e.g. environmental flows).
- In cases of extreme weather events that go beyond the scope of risk management tools, a set of pre-defined crisis management procedures is needed to mitigate the costs of water shortages or excesses.

The last pillar for an efficient policy approach to managing droughts and floods is insurance and compensation systems. For several reasons, crop insurance markets against catastrophic risks, especially for droughts, have rarely been developed in OECD countries without significant government support (notably through crop insurance subsidies and/or public reinsurance of last resort). While government intervention can be justified for catastrophic risk layers, an efficient insurance system should be based on risk-based premiums to provide incentives that encourage production choices that limit exposure and vulnerability of farm systems to droughts and floods. Hence, a clear definition of boundaries between layers of risks is a prerequisite for sound government intervention. In some contexts, innovative tools such as catastrophic bonds and weather index-based insurance could also be developed to complete the existing set of market and policy responses. If necessary, government support to agricultural insurance should be limited to avoid artificially boosting demand for insurance beyond the primary motive of improving risk allocation.

Chapter 1

Characterising and measuring droughts and floods

This chapter provides an overview of the meteorological, hydrological and socio-economic dimensions of drought and flood risks. It describes the approaches to characterise and assess these risks, and to measure their costs to agriculture and other sectors. It serves as a background for the economic and policy analysis developed in subsequent chapters.

1.1. The meteorological, hydrological and socio-economic dimensions of drought and flood events

One can consider three dimensions to characterise drought and flood events: meteorological, hydrological, and socio-economic, with each having a distinct set of indicators (for more details, see Box 1.1). Although distinct, these dimensions are nevertheless related (Figure 1.1). The most common indicators used are meteorological and hydrological indicators, as economic, social and environmental impacts are more difficult to assess and can vary a lot across affected people and regions. However, measuring impacts is important to assess the costs and benefits of mitigation strategies against flood and drought risks.[1]

At the source of droughts and floods, there is always a weather event or set of combined weather events. The meteorological dimension of droughts and floods basically focuses on, respectively, the deficit or excess of precipitation compared to reference values. Precipitation, either in excess or in deficit, is the main factor, but other variables such as temperature, humidity, air and soil, and wind also play a role. For example, in the case of the summer melting of glaciers that feed rivers downstream, temperature is a major factor driving river flows and potential flooding risk. Basic indicators of the meteorological dimension include, for example, percentage of normal precipitation, which is calculated as a ratio between current precipitation and a past historical average for a given period of time, but also with more complex indicators such as the Standard Precipitation Index.[2] Other characteristics are important to characterise meteorological floods and droughts, notably their time of occurrence, duration and spatial characteristics (location and spatial extent). Most often, floods relate to sudden events resulting from windstorms or heavy rain while droughts are considered as cumulative events over an extended period of time, from a few weeks to several months or even years. Drought characteristics vary a great deal across continents, countries, and regions, as illustrated by Figure 1.2 which presents the number and duration of major drought events between 1950 and 2000 at the continent level, with implications for the appropriate management of these risks.

Box 1.1. Defining droughts and floods

There are a large number of definitions of droughts and floods, from more abstract to more concrete, from more descriptive to more operational. Droughts and floods are indeed an object of study in a variety of disciplines such as meteorology, hydrology, agronomy, economics, sociology, psychology, and political science among others. In practice, water managers and policy makers need to rely on working definitions to plan their programmes and pilot their interventions in the course of action. Economists aim at measuring the costs of droughts and floods, which requires a different approach and eventually different indicators. Sociologists focus on topics such as the collective response to extreme events, such as crisis management. Definitions may therefore vary depending on objectives, local conditions and socio-economic contexts. There is no "one size fits all" definition of either droughts or floods.

Notwithstanding these complexities, droughts can be broadly defined as a temporary decrease of water availability in a given water system, caused by prolonged deviations from average levels precipitation. Drought essentially "differs from other natural disasters in the slowness of onset and its usual lengthy duration", which makes it difficult to determine the onset and duration of a drought event (European Commission, 2007; Wilhite, 2007). Drought is a normal and recurrent feature of climate, although it can evolve under certain circumstances into a disaster, depending mainly on the vulnerability of the affected society and its capacity to manage the impacts, and on the severity and duration of the event (Kampragou et al., 2011).

From a general point of view, floods can be defined as "rises, usually brief, in the water level of a stream or water body to a peak from which the water level recedes at a slower rate" (WMO and UNESCO, 2012). The main types of flood risk that affect agriculture are river floods, flash floods, and coastal floods. River flooding occurs when the river capacity system is insufficient to contain the flow of water in the river. Flash floods "arise from intense, localised rainfall, and can happen practically anywhere" (World Bank, 2010). Coastal zones are "subject to flooding as a result of storm surge-increased sea levels driven by tropical storms or by strong windstorms arising from intense offshore low-pressure systems" (World Bank, 2010).The impact of flooding on agriculture varies with crops' tolerance or land use activity, and the characteristics of the flood event (frequency, duration, depth and seasonality) (Morris et al., 2010).

Figure 1.1. Characteristics and impact of droughts and floods

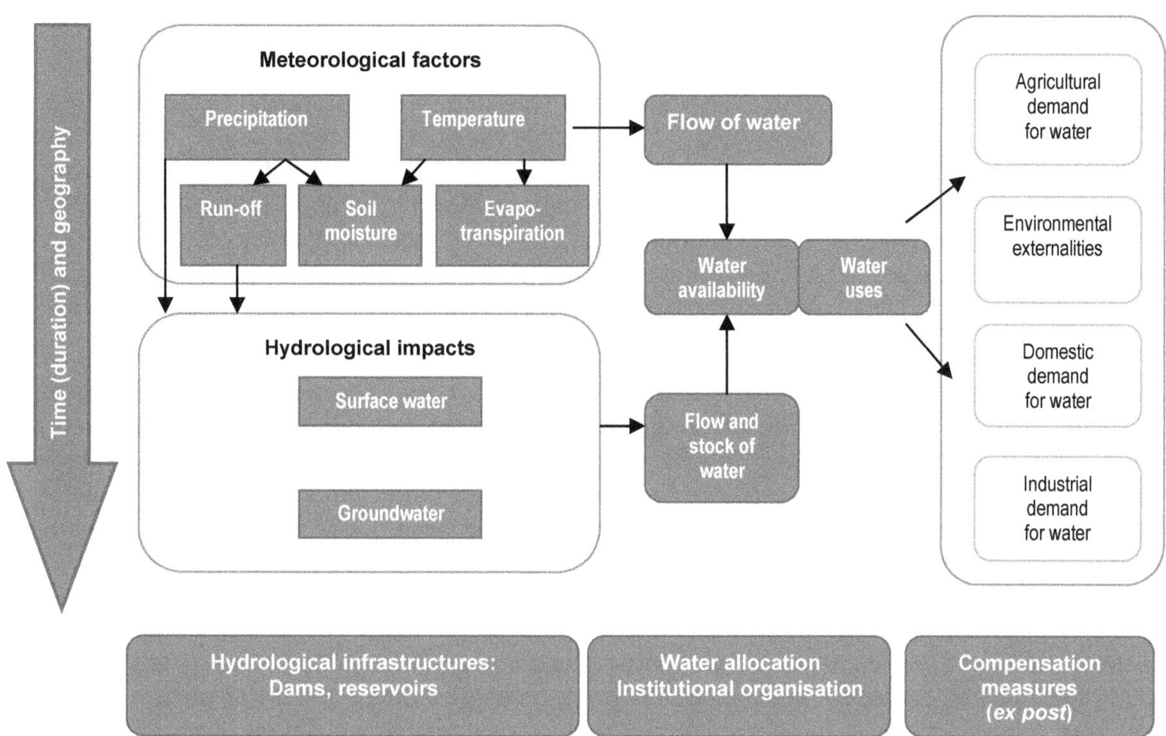

Figure 1.2. Number and duration of droughts in the different continents

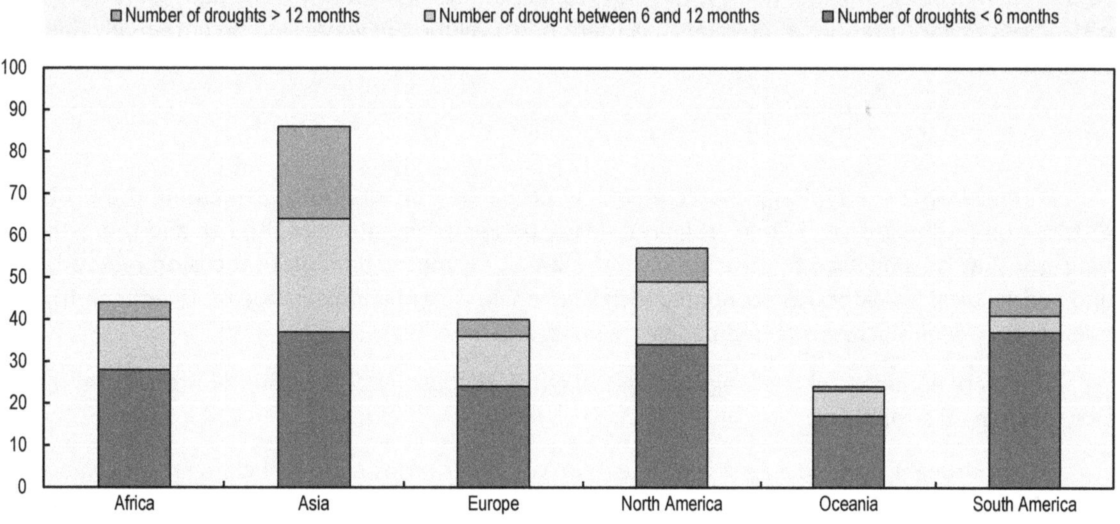

Source: Based on Sheffield, J. et al. (2009).

The hydrological dimension of floods and droughts refers to the state of water resources in the different compartment of the water system. Basically, excess or deficit rainfall is transmitted to the different components: soils, surface water and groundwater. This transmission may have some time dependency and inertia; for instance, affecting soils first, then surface water, and lastly groundwater. The overall water availability for water users depends on the status of the stock and water flow in the different compartments of the water system. Measuring the state of hydrological systems can be done through the use of indicators: soil moisture, river flows, and levels of groundwater, lakes and dams. The state of water resources is not solely driven by meteorological conditions; rather it is the joint outcome of weather conditions and anthropic use of water: e.g. management of river banks, reservoirs, dams, land cover, and drainage which influence how deficit or excess rainfall is transmitted to hydrological compartments. Water systems are not just ecosystems, but socio-ecosystems.

These considerations lead to an important distinction. While floods are usually defined from a hydrological perspective, there is a net distinction between droughts, a meteorological notion, and water shortages, which relate to a short-run deficit between aggregate water use and aggregate water availability in a given water system. This means that not all droughts translate into water shortage events since it depends on the exposure and vulnerability of affected societies. Finally, water shortages should be distinguished from water scarcity, which corresponds to an aggregate deficit between water use and availability, and in the longer-term, to a structural perspective.

The economic, social and environmental dimensions of droughts and floods relates to the economic, social and environmental impacts of these extreme events. Economic impacts of droughts include direct damages faced by sectors that heavily rely on water for their production process such as agriculture, hydroelectricity, water intensive manufacturing and households. Floods most often affect the productive assets of households and different industries, with potential longer lasting negative impacts on production capacities. Droughts and floods may also have social implications, notably when they affect poorer and more vulnerable categories of the population. Finally, the environmental impact of droughts and floods can also affect the associated ecosystem services and thus social welfare. Examples of this include an increase in the erosion transfer of sediments and nutrients, which results in a decrease of water quality; the non-respect of minimum water flows during drought events can increase concentrations of pollutants as there is less dilution in water and be potentially damaging to aquatic life.

1.2. Assessing and characterising drought and flood risks

Formally, risks are typically characterised by probability distributions. Estimating the probability distribution of droughts and floods requires clearly defining the nature of the risk and the associated indicator that is being used to measure it. In this area, it is more common to focus on meteorological and hydrological indicators, as economic, social and environmental impacts are more difficult to assess and can vary widely across affected people, regions, and time.

There are a number of challenges in assessing the probability distribution of droughts and floods risk, including the following:

- Each drought or flood is to a certain extent unique in terms of spatial characteristics, intensity, impacts, etc. Thus estimating a risk distribution requires some standardisation of definitions which may be challenging in practice.
- Non-stationarity of exposure and vulnerability. Climate change, but also risk exposure, is non stationary: farmers change their cropping patterns over time; the composition of land cover is evolving between urban areas, agriculture, forests and other areas. This makes it difficult to

build a stable relation between a given meteorological event and its associated impact on human systems.

- For "catastrophic risks" or "extreme events", there are specific methodological challenges to risk assessment which makes it difficult to have reliable risk estimates, compared to more common or frequent weather events (Box 1.2).

Box 1.2. Assessing the probability of extreme weather events: A statistical challenge

Assessing the probability of extreme events is a statistical challenge. It is possible to use descriptive statistics, however the observed period of rainfall disasters might not include all possible extreme values because some critical droughts and floods events are so rare that they are not observed in the data sample. Extreme Value Theory (EVT) is one way to address this issue by providing estimations of the occurrence of values beyond the observed extremes (Reiss and Thomas, 2007). The principle of EVT is to calibrate a probability distribution on the flow of water (precipitation, discharge of water on river) and extrapolates the distribution function beyond the observed extremes. The extrapolation results can be used to analyse the probability of occurrence of events like droughts and floods that could not otherwise be observed. This approach has, however, limitations (Annex 1.A1).

Such methodology leads to an estimation of the frequency of an event of a given *size*. Usually these frequencies are expressed as return periods of extremes (or average waiting times between the occurrences of extremes of a fixed size). For example, a return period of a 100–year flood is defined as a flood that can occur on average once every 100 years. Although very long periods return values can be calculated from the fitted distribution, the confidence that can be placed in the results is reduced for very long return periods. Additionally, there are potentially large biases due to uncertainty in a climate change context.

In sum, in spite of specific statistical tools on the estimation of extremes, the management of extreme weather events such as droughts and floods must deal with considerable residual uncertainty. Non-stationary climate due to global warming adds another layer of uncertainty to this global picture. A change in the overall distribution of precipitation modifies the value of average precipitation. A shift in the precipitation distribution can greatly increase the likelihood of extreme values above the critical threshold on the left tail of the distribution. This means that the effect of a shift to the left implies less precipitation on average and more situations with extreme deficit* of precipitation.

* "Deficit" means a level below a physical threshold, while the word "shortage" will be used to refer to a level below demand.

Another important issue associated with the definition and characterisation of droughts and floods is that of the "extreme", "catastrophic" or "disastrous" nature of these risks, especially as this may have implications in terms of policy responses to their management. While these three terms seem to be closely related, they are in fact very different. IPCC (2012) defines an "extreme weather event" as "the occurrence of a weather variable above (or below) a threshold value near the upper (or lower) ends of the range of observed values and variables". This is essentially a statistical perspective, for which the range of extreme values in the probability distribution is respectively represented by the extreme right and left tails.

Such a statistical approach also means that extreme weather events are relative, based on a benchmark of what are considered as average or "normal" weather conditions. It often means that extreme weather events are characterised by low probabilities; however this ultimately depends on the overall shape of the probability distribution function. In some countries, extreme weather events can occur frequently, every five or even three years depending on the region considered. In these cases, it is inappropriate to speak about "extreme" weather events in that sense. Examples include severe droughts in Australia and Spain, which are considered as common weather conditions.[3]

While extreme weather events can be defined in an objective way, the notions of catastrophes and disasters are more subjective and convey very different meanings. Firstly, compared to the notion of extreme weather events, notions of "catastrophes" and "disasters" place more emphasis on the economic, social and environmental impact of climate events rather than their probability and meteorological features. Disasters are often measured in terms of human losses and monetary costs, or loss of biodiversity. The two concepts can coincide: extreme weather events are typically more likely to cause catastrophic economic, social and environmental damages, but this is not necessarily the case. The extent of the damage depends on the exposure and vulnerability of the affected system, which can vary across regions and depend on policy responses to manage droughts and floods. It can be also the case that poor management of moderate drought or flood events may result in catastrophic damages.

Secondly, the terms "catastrophes" or "disasters" have a strong subjective and emotional content, and are used in varying contexts. The term catastrophe or disaster can be applied equally well to a plane crash, an epidemic or to a poor harvest due to drought. Going a step further, one could say that the use of the word catastrophe is more of a prescriptive than descriptive nature. This can be illustrated in practice by the fact that in a number of countries, declaring an area in a state of natural disaster involves triggering exceptional specific responses, beyond the set of usual risk management tools. Calling an event a catastrophe is thus an act that defines the boundaries of risk management.

1.3. Assessing the costs of droughts and floods to agriculture and other sectors

Economic impacts caused by drought affect agriculture through losses in crop yields and livestock production, but also through increased insect infestations, plant diseases and wind erosion (European Commission, 2007). Droughts are rarely considered as a catastrophic event for the agricultural sector in OECD countries at the macroeconomic scale, although they can significantly affect the agriculture sector. Flooding is a significant hazard worldwide, with an estimated USD 700 billion of damages in the period 1985-2008 (Morris et al., 2010). Agriculture, however, by occupying a large proportion of the landscape, can be at once affected by flood risk, but also play a positive role in the overall mitigation of flood risks in water systems as a provider of floodplain areas, or through the influence of upstream agronomic practices reducing flood risk downstream (Morris et al., 2010).

Economic costs of droughts and floods in agriculture can be direct or indirect, instantaneous or induced (Table 1.1). Direct and short-run costs typically include crop and livestock production losses. Direct but induced (longer term) impacts include losses of productive assets such as machinery and buildings, and declines in land value, which can reduce productivity in the longer run and may require significant resources for recovery. Indirect impacts can also be born outside the agriculture sector by related sectors or through agricultural markets. For example, a severe drought in a large commodity exporting country may cause a substantial drop in global agricultural production, leading to a rise in commodity prices at the world level. When insurance markets or compensation mechanisms are in place, the cost of droughts and floods can also be borne at a broader level by insurance policyholders or taxpayers depending on the risk sharing characteristics of these mechanisms.

Typically, direct and tangible costs, i.e. reduction in crop yield, are easier to quantify than are indirect, non-tangible costs such as loss in biodiversity or soil erosion and its associated impact on soil productivity (Volker et al., 2012; Meyer et al., 2013). In some cases, direct tangible costs of droughts and floods may indeed just represent the "tip of the iceberg" as compared to the "real" (but unobserved) full cost of these weather events. Such heterogeneity in cost assessment is susceptible not only to lead to underestimation of the costs of droughts and floods, but also to be a source of bias in favour of sectors for which the cost estimates are more feasible. Caution is therefore necessary when it comes to comparing the costs of droughts and floods between countries and across time.

Another challenge is that the agriculture sector faces a variety of risks that can affect production and income: weather risks, pests and diseases, price risks (on both input and output sides), institutional risks (due to changes in regulations or policy environment). The relative importance and combination of these risks can vary a great deal across countries and farming systems, which can combination of these risks can vary a great deal across countries and farming systems, which can explain different kinds and degrees of policy responses to address them (OECD, 2011). The impact of droughts and floods on farmers' income should be considered thus holistically. Risks can be positively correlated; for example, drought or flood conditions may increase the risk of pests and diseases in some cases, and thereby aggravate crop losses. However, one can observe in some countries a negative correlation between yield and price risks which can, to a certain extent, offset each other in the formation of farmers 'incomes (OECD, 2011). In certain cases, such income can increase if the price swing more than compensates yield losses, and risk is transferred through prices in the value-chain of agricultural commodities.

Without a systematic and harmonised method to assess the costs of droughts and floods in agriculture, one has to rely on case-by-case evaluations. For example, the 1976 drought in France reduced farmers' income by 9% compared to more typical years; the 2005 drought lead to a 22% decrease in farmers' income compared to 2004 (Amigues et al., 2006). Drought is quantitatively the most important risk in France, representing on average 57% of indemnities paid by the French National Guarantee Fund for Agricultural Disaster Risks (Babusiaux, 2000). For some of the most important droughts of the last 40 years, price increases have in a few cases slightly offset yield losses, but external drivers of commodity prices seemed to have been more important drivers than the production failure itself.

Table 1.1. Cost of floods and droughts to agriculture sector

Floods		
	Instantaneous (short-run)	**Induced (medium to long-run)**
Direct	Human fatalities Damage/destruction of economic goods Emergency costs Fatalities to livestock Damage/destruction of infrastructure	Loss of value added due to damages on production factors Rehousing of households Relocation of livestock
Indirect	Increase in travel time due to damage on infrastructure Delay or cancellation of supply from the flooded area (e.g. inputs, machinery)	Loss of value added due to business interruption of assets in the flooded area Loss of value added due to damage on infrastructure (Re)financing costs, borrowing costs
Droughts		
Direct	Damage/destruction of economic goods Negative impacts to livestock production and health Reduced yield and crop acreage for agriculture due to drought related water shortage	Loss of value added due to damages on production factors, e.g. soil Relocation of livestock
Indirect	Higher cost or irrigation water pumping or water prices Cost of buying additional external feed due to reduced pasture production Increasing cost of farming operations due to inappropriate soil conditions or excessive heat	Loss of value added due to business interruption of assets Loss of value added due to damage on infrastructure (Re)financing costs, borrowing costs

Source: Based on Brémond, P., F. Grelot, and A.-L. Agenais (2013). www.nat-hazards-earth-syst-sci.net/13/2493/2013/nhess-13-2493-2013.pdf.

Figure 1.3. Crop insurance indemnities and disaster assistance payments in the United States, 1970-2013

Billion 2013 USD

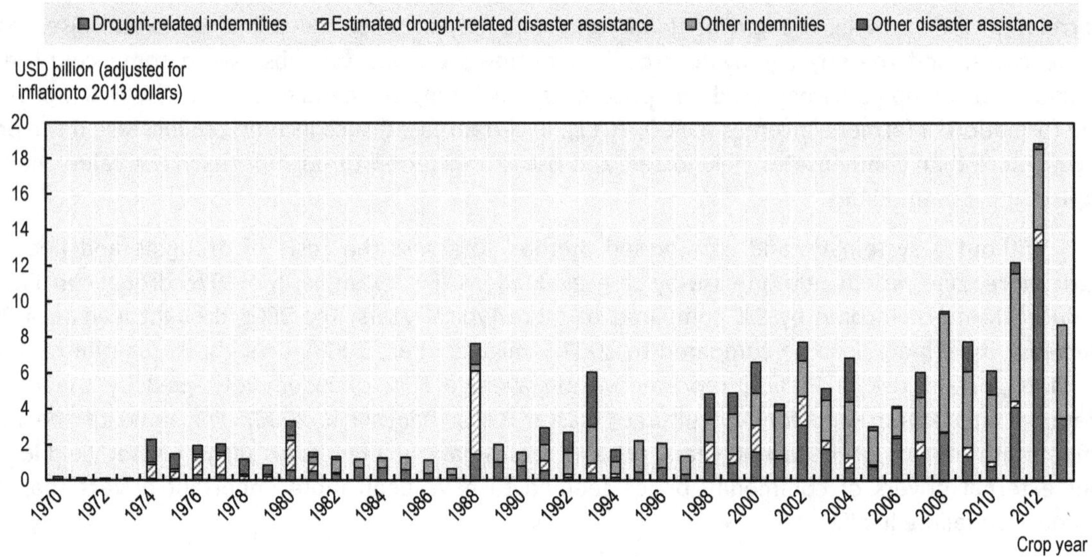

Source: Wallander et al. (2013), www.ers.usda.gov/media/1094660/err148.pdf.

In the United States, drought is the primary risk for agricultural production, although exposure can vary a great deal across states. On average, drought risk has been estimated to represent about 40% of crop insurance indemnity payments made between 1948 and 2010 (Wallander et al., 2013). Figure 1.3 presents the crop insurance indemnities and disaster assistance payments for droughts and other weather events since 1970, illustrating the importance of this risk. The 2011-2012 droughts in United States led to unprecedented levels of compensation compared to previous years, although the increasing scale of the programme also explains their upward trend. The 2014 drought in California also had large costs for agriculture and other water users and uses. In terms of price effects, one can observe a significant negative price-yield correlation, especially in the Corn Belt, providing a partial natural hedge for farmers' revenues (Harwood et al., 1999).

A lack of comprehensive knowledge of the costs of droughts and floods is a barrier to the improvement of policy approaches to managing these. In particular, it does not allow to undertake sound cost-efficiency or cost-benefit analysis that could provide useful guidance to citizens, stakeholders and decision-makers. The lack of data and knowledge is not the single barrier: even from a methodological perspective, cost-benefit analysis is more complex when introducing risk and uncertainty. Lack of data on costs is not only an issue for *ex ante* assessment, but also can be used as decisive arguments in the course of action. For example, in a negotiation process for water restrictions between different waters users (agriculture, industry, tourism) and uses (environmental flows), using cost figures in monetary terms is a powerful rhetorical tool during the bargaining process; however, it may ultimately lead to inefficient water allocation or favouring short-term mitigation efforts over long-run sustainability.

In view of these methodological and data limitations, some encouraging progress is being made. For example, the European project *Cost of Natural Hazards* (Volker et al., 2012) funded by the EU *Seventh Framework Programme for Research and Technological Development* (FP7, 2007-2013) undertaken between 2010 and 2012, conducted a state-of-the-art assessment of costs of natural hazards such as droughts and floods. It also analysed their appropriate use in cost-benefit analysis of mitigation and prevention strategies, and identified the main data and methodological gaps (Meyer et al., 2013). The scope was not limited to agriculture but all sectors concerned by natural hazards. The report proposed a comprehensive set of recommendations to expand and improve the assessment of the costs of natural hazards, including specific advice on droughts and floods with a view to better integrate this information in decision-making.

Finally, there are linkages between the meteorological and hydrological dimensions of floods and droughts and their impact on water users and uses, which can be measured by the effects of the water shortage or excess in terms of financial and economic losses to farmers.[4] The physical characteristics of droughts and floods, such as the time of occurrence, duration and spatial extent, have strong implications on the economy. Timing is important: droughts or floods that occur outside a critical vegetation period will have less economic impact. The spatial scale is important to evaluate agricultural impacts as some agents may be adversely affected as a result of spatial correlation. Spatial correlation risk exists when extreme weather events affect large geographical locations at the same time, causing wide-scale damage to agricultural production. The spatial extent and severity of droughts will vary seasonally and annually, whereas the spatial extent of floods is more predictable when using topographic information.

Notes

1. It is not the purpose of the present report to make an extensive review of the definitions of droughts and floods, but to focus on the elements that are meaningful for good policy approaches to manage, mitigate and cope with these risks in the short and long term.
2. The Standard Precipitation Index measures the relative rarity of a drought event in terms of cumulative rainfall at a given location, for a given period of time. It is adjusted statistically using the gamma distribution.
3. At the polar case, when drought conditions are permanent, one speaks of aridity.
4. Correlations between weather events and damages are often difficult to model, except for the most catastrophic events. Assessments of damages caused by floods of short duration, such as damages to field crops, can be undertaken immediately by a physical inspection. In the case of droughts, a set of statistical studies undertaken in the 2000s analysed the influence of climatic parameters — rainfall, temperature, soil moisture — on crop yields. They showed that a limited number of climatic variables, especially the sum of degree-days, are able to explain a significant proportion of crop yields. In addition, they identified thresholds beyond which crop yields decline drastically (Ortiz-Bobea, 2013; Roberts et al., 2013). These studies are mainly used to project the impact of climate change on crop yields, rather than for a practical estimation of crop losses for, notably, indemnification purposes.

References

Amigues, J.P., et al. (Ed.) (2006), *Sécheresse et agriculture. Réduire la vulnérabilité de l'agriculture à un risque accru de manque d'eau*, Expertise scientifique collective, synthèse du rapport, INRA, France.

Babusiaux, C. (2000), *L'assurance récolte et la protection contre les risques en agriculture*, Ministère de l'Agriculture et de la Pêche, Ministère de l'Économie, des Finances et de l'Industrie, Paris.

Brémond, P., F. Grelot, and A.-L. Agenais (2013), "Review article: Economic evaluation of flood damage to agriculture – review and analysis of existing methods", *Natural Hazards and Earth System Sciences*, Vol. 13, pp. 2493–2512, www.nat-hazards-earth-syst-sci.net/13/2493/2013/nhess-13-2493-2013.pdf.

European Commission (2007), "Mediterranean water scarcity and drought report. Technical report on water scarcity and drought management in the Mediterranean and the Water Framework Directive" *Technical Report - 009 – 2007*, Mediterranean Water Scarcity and Drought Working Group, Brussels.

Harwood, J., R. Heifner, K. Coble, J. Perry, and A. Somwaru (1999), "Managing Risk in Farming: Concepts, Research, and Analysis", *Agricultural Economic Report*, No. 774.

IPCC (2012), *Managing the Risks of Extreme Events and Disasters to Advance Climate Change Adaptation. A Special Report of Working Groups I and II of the Intergovernmental Panel on Climate Change*, Cambridge University Press, Cambridge and New York.

Kampragou, E., S. Apostolaki, E. Manoli, J. Froebrich and D. Assimacopoulos (2011), "Towards the harmonization of water-related policies for managing drought risks across the EU", *Environmental Science and Policy*, Vol. 14, pp. 815-824.

Klein Tank, A.M.G., F.W. Zwiers and X. Zhang (2009), "Guidelines on analysis of extremes in a changing climate in support of informed decisions for adaptation", *Climate data and monitoring* WCDMP-No. 72, WMO-TD No. 1500.

Meyer, V. et al. (2013), "Review article: Assessing the costs of natural hazards – state of the art and knowledge gaps", *Natural Hazards and Earth System Sciences*, Vol. 13, pp. 1351–1373, www.nat-hazards-earth-syst-sci.net/13/1351/2013/nhess-13-1351-2013.html.

Meyer, V. et al. (2012), *Costs of Natural Hazards – A Synthesis,* Costs of Natural Hazards – A Synthesis, available at: www.ufz.de/export/data/2/122169_CONHAZ_WP09_1_Synthesis_Report_final.pdf.

Morris, J., T. Hess and H. Posthumus (2010), "Agriculture's Role in Flood Adaptation and Mitigation: Policy Issues and Approaches", in OECD, *Sustainable Management of Water Resources in Agriculture*, OECD Publishing, Paris. http://dx.doi.org/10.1787/9789264083578-9-en.

Roberts, M.J., W. Schlenker and J. Eyer (2013), "Agronomic Weather Measures in Econometric Models of Crop Yield with Implications for Climate Change," *American Journal of Agricultural Economics*, Vol. 95, No. 2, pp. 236-243.

Sheffield, J., K.M. Andreadis, E.F. Wood, and D.P. Lettenmaier (2009), "Global and continental drought in the second half of the 20th century: severity-area-duration analysis and temporal variability of large-scale events", *Journal of Climate*, Vol. 22(8), pp. 1962-1981.

Volker M., N. Becker, M. Markantonis et al. (2012), *Costs of Natural Hazards – A Synthesis,* Costs of Natural Hazards – A Synthesis, available at: http://conhaz.org/outcomes-1.

Wallander, S., M. Aillery, D. Hellerstein, and M. Hand (2013), *The Role of Conservation Programs in Drought Risk Adaptation*, ERR-148, US Department of Agriculture, Economic Research Service, Washington, DC. www.ers.usda.gov/media/1094660/err148.pdf.

Wilhite D. (2007), "Preparedness and Coping Strategies for Agricultural Drought Risk Management: Recent Progress and Trends", In: Sivakumar, M.V.K. and R.P. Motha, (eds.) *Managing Weather and Climate Risks in Agriculture*, Springer, London.

World Bank (2010), "Assessment of Innovative Approaches for Flood Risk Management and Financing in Agriculture", *Agriculture and Rural Development Discussion Paper 46*, Washington DC.

World Meteorological Organization and UNESCO (2012), *International Glossary of Hydrology*, WMO Publishing, Geneva.

Annex 1.A1

Statistical theory of extreme values

The Peak Over Threshold and Block Maxima approaches are two methodologies derived from the extreme values theory to assess probabilities of extreme events. They are different by the way they select extreme events in the data sample and by the distribution probability which is applied according to their separate methodology.

The *Block Maxima approach* requires that the maximum value observed in each time span (time span is arbitrary selected, e.g. weeks, months) be selected. This approach consists of representing the extreme values by the Generalised Extreme Value (GEV) distribution (combination of the Gumbel, Fréchet, and Weibull maximum extreme value distributions). The *Peak Over Threshold (*POT*)* method considers values greater than a defined threshold and is based on the Generalized Pareto Distribution (GPD) distribution (Reiss and Thomas, 2007).

The main drawback of the *maxima* approach is that events are selected for every single year of measurement. Peak over threshold selects extreme values for every value above a defined threshold and it has proven to be more flexible than block maxima methods. However, the choice of threshold in selecting extreme values is an important practical problem, which is based on a trade-off between bias and variance. The threshold must at once be high enough for the excess over the threshold to follow GPD and allow for a large enough sample size (Reiss and Thomas, 2007; World Meteorological Organization, 2009).

Both methodologies require the assumption of independent and identically distributed events, e.g. the sampling does not select two "nearby" maxima which relate to the same larger flood mechanism. But many environmental variables (temperatures, precipitation, wind speed) are temporally correlated and there are seasonal and long term trends. The assumption of independence can be relaxed by dealing with cluster maxima instead of all exceedances in the POT method, and one way to relax the assumption of identical distribution is to allow the parameters of the Poisson–GP model depend on covariates (e.g. annual or diurnal cycles).

The extrapolation or forecasting of extreme values techniques relies on imposing a probability distribution law and then inverting this distribution to calculate frequency (Reiss and Thomas, 2007).

Chapter 2

The economics of droughts and floods in agriculture

This chapter presents and analyses market failures related to water-related risks and their implications for public policies. It also presents the main policy and market drivers that affect the vulnerability of agriculture to droughts and floods.

2.1. The economics of risks in agriculture

The fundamental starting point of an economic approach to risks is that they impose certain costs to economic agents, which can in principle possible be expressed in monetary terms. This idea is conceptually formalised by the notion of risk premium, defined as the willingness to pay in monetary terms of a given economic agent to transform a given risk into a certain outcome equal to the expected outcome of the risk itself.[1] In economics, the cost of risk depends essentially on:

- The characteristics of the risk, which is described by a set of probabilities and associated outcomes (or probability distribution functions).
- The agents' preferences towards risks. The economics of risk generally synthesises such preferences under the concept of risk aversion. Basically, the more an agent is risk averse, the more he is willing to pay to eliminate a given risk, all things being equal.

It is generally assumed that most economic agents are risk averse, although the degree of risk aversion can vary a great deal. A typical assumption also made is that the degree of risk aversion tends to decrease with the level of "wealth" of the economic agent: as individuals become richer, their willingness to pay to eliminate a given risk decreases. In agriculture, due to the intrinsically risky nature of the economic activity, numerous studies have attempted to estimate farmers' degrees of risk aversion and to characterise the main explanatory variables (OECD, 2011). Knowing farmers' risk preferences is useful for assessing the costs and benefits of risk management tools and public policies devoted to agricultural risk management. Eliciting farmers' preferences towards risks nevertheless remains a challenging task, in particular to disentangle pure risk preferences from technological considerations. On the basis of the extensive literature in this area in many OECD and non-OECD countries, a reasonable order of magnitude for the coefficient of farmers' risk aversion is included in a range between 2 and 7-8.

From the standard microeconomic perspective, it is important to note that risk aversion and the associated risk premium are essentially psychological notions that can be understood as specific tastes of agents towards risks. But beyond the psychological cost of risk, there are also cases in which risks can reduce expected outcomes, such as expected farmers' profits, and thus represent another form of cost to farmers should they be risk neutral as regards their risk preferences. This is notably the case when the income tax schedule is convex or when capital market imperfections make the cost of refinancing after a negative shock both higher and convex (Froot et al., 1993).

Because risks are costly to farmers, they have an incentive to invest in risk management activities, including risk mitigation, risk sharing, and risk transfer. Risk mitigation relates to costly activities that reduce the impact of risks for the farm. One usually makes a distinction between self-insurance, which consists in reducing the loss of a given risk when it occurs, and self-protection, which consists in reducing the probability of the risk (Ehrlich and Becker, 1976). Risk management activities are diverse in agriculture. Typical examples include the use of drought-resistant crop varieties, crop diversification, and the use of pesticides. Farmers often face a large array of risks: weather, pests, diseases, etc., which they can mitigate through their choices of agronomic practices. Most often there is no free lunch: risk mitigation activities involve a trade-off between expected outcomes (e.g. income) and the variability of these outcomes.[2]

In addition to risk mitigation, farmers can also reduce the cost of risk by sharing risks with other farmers though insurance markets, or transferring risks to investors through the use of financial instruments. The insurance mechanism is based on the law of large numbers, which allows to spread risks among the insured in exchange for an insurance premium which is equal to the expected cost of the risk (notwithstanding the administrative costs of running the insurance system itself). Risk sharing through insurance markets can, in principle, lead to an economically efficient allocation of risks among economic agents. In practice, several problems, such as market or government failures, hinder the proper development of insurance markets in agriculture. These include: asymmetric information (adverse selection, *ex ante* and *ex post* moral hazard), ambiguity aversion, and lack of knowledge of the risks. These problems are especially important in the case of insurance against droughts and floods in agriculture.

Finally, risk management is closely related to the question of time. The time dimension of risk management is often under-considered. Firstly, farmers make investment choices, which can be capital intensive, and put them on a specific development path over a certain time horizon. These investment strategies may have consequences on risk exposure over the medium or short term. If a farm specialises in a narrow set of crops, risks related to climatic or market conditions of these crops can change and it may be very costly to revise the whole farming system to adapt to these changing circumstances. This is not specific to agriculture. Secondly, risks can be managed using the time dimension by building precautionary savings to smooth income across time and reduce short-term vulnerability arising from negative income shocks due to climate or markets.

Among the large set of risks farmers commonly face, drought and flood risks have several specificities that have implications in terms of policy responses by governments, and that will be described more precisely in the following sections:

- Water is a special "economic good" with specific features, which has implications for the mitigation of water-related risks that are droughts and floods;
- Insurance markets against drought and flood risks are often incomplete due to a complex set of behavioural, market, and government failures.

2.2. Market failures related to water-related risks

Water is an essential input for agricultural production, including rain-fed and irrigated crops as well as livestock, whether directly or indirectly. In the case of rain-fed crops and pastures, soil moisture is a key factor in the production process, and is mainly dependant on weather conditions — although farmers can through certain agronomic practices influence soil moisture storage. For rain-fed agriculture, water is not an input in the traditional economic sense meaning that farmers have to buy it, but rather an exogenous production factor. In the case of irrigation, farmers rely on water resources from different water compartments (surface and groundwater), which require specific equipment. Access to irrigation water depends on the physical characteristics of the water system and weather conditions (precipitation, evapotranspiration), which are exogenous, but also on institutional rules that define rights to use water and the associated costs of doing so.

Water resources are used for many different purposes, ranging from extractive uses such as irrigation, to non-extractive uses, for example recreation or the maintenance of well-functioning ecosystems (OECD, 2010). These different types of water uses are associated with positive economic values that can be either private (e.g. irrigation water) or collective (e.g. minimum environmental flows). Water uses can be rival, i.e. there is competition across a finite water resource stock. However non-extractive uses (requiring no withdrawal or diversion from ground or surface water sources) and non-consumptive uses (that return flows to the system) allow the re-use of water by other users, which tends to mitigate the assumption of rivalry.

A second important dimension to characterise water resources from an economic perspective is their degree of excludability. Excludability refers to the possibility of a given community of water users to control water withdrawals in a given water system, including the ability to exclude some users from the community. Excludability depends on many different factors: the characteristics of the water system (surface or ground water, complexity of the water system), the costs of control, the credibility of penalties in case of unauthorised withdrawals, etc. It is not just a physical characteristic of the water resource, but includes an economic dimension (excludability requires a set of rules and actions that have costs and benefits). If, for several of these reasons, excludability is low, then there is a risk of free-riding, with potentially inefficient water allocation.

From an economic perspective, the combination of the characteristics of rivalry and excludability categorise water resources, with implications in terms of policy approaches to managing them. Table 2.1 summarises four stylised types of goods depending on two characteristics: rivalry and excludability (Ostrom et al., 2003; Ostrom, 2005). Goods that are either non-rival, or rival but non-excludable, are typically defined as "collective goods" (Starret, 2003). These include club goods (high excludability, low rivalry), pure public goods (low excludability and rivalry), and common pool resources (low excludability, high rivalry).

Table 2.1. Categories of goods depending on the degrees of rivalry and excludability

		Rivalry	
		High	Low
Excludability	High	Private Good	Toll/Club Good
	Low	Common Pool Resource	Pure public Good

Source: Ostrom (2005), http://press.princeton.edu/chapters/s8085.pdf.

As mentioned above, the degrees of rivalry and excludability can vary across uses, circumstances, location, time, and so no categorisation should be seen as "etched in stone."[3] In particular, the degrees of rivalry and excludability may also vary with water availability. For most uses, water is typically assumed to exhibit a positive and decreasing marginal value up to a certain level, beyond which there is water satiety (i.e. the marginal value of water is null). Beyond an upper threshold, water has a negative marginal value: excess water causing crop damages, losses due to excess water or floods (Table 2.2).

Table 2.2. Water as an economic good in drought and flood circumstances

Water availability	Drought	Moderately dry	Moderately wet	Flood
Rivalry	Very high	High	Low -medium	Highly negative
Excludability	High if droughts are frequent Depends on hydrological characteristics and infrastructures	Medium to high if frequent water deficits Depends on hydrological characteristics and infrastructures	Low Depends on hydrological characteristics and infrastructures	Low to medium Depends on hydrological characteristics and infrastructures
Type of good	Private good	Private good Toll / club good	Common pool Public good	Common bad

In situations of water shortage due to droughts, the degree of rivalry, and the associated opportunity costs of water, is likely to be exacerbated, especially if marginal values of water are heterogeneous across water users. In the case of floods, the marginal value of water is negative, hence water is considered as a "bad". There is some form of rivalry to avoid damages from water excess, and one can reasonably assume that the degree of this rivalry is also likely to increase with the level of this water excess. For example, floodplains can allow mitigating flooding in urban areas up to certain limits, beyond which the water storage capacity of floodplains will no longer be able to make it. Flood-control or flood-protection infrastructures, such as dams or floodplains, are usually considered as public or common goods that reduce flood risks in a given area of the watershed.

The way water quantity affects excludability is not so straightforward. The degree of excludability of a water resource depends on a large set of factors such as the type of water resource considered (e.g. surface water or groundwater), available technologies to measure water consumption, and monitoring tools. Excludability also depends on institutions and can sometimes be enforced outside a given community but not within its members. From an economic perspective, ensuring or increasing excludability of given water resource has costs and benefits. The benefits of excludability mainly arise from a more socially efficient allocation of water and are thus susceptible to increase with rivalry. A low level of excludability might be cost-effective in water basins where water is abundant compared to demand, while in basins with frequent shortages due to droughts, a higher degree of excludability (through water metering or monitoring) would be required by cost-benefit considerations.

This categorisation has important normative implications in terms of policy responses. Indeed, common pool resources typically suffer from overconsumption and inefficient resource allocation due to the lack of excludability. This is a well-known case of market failure regarding water resources in general, which calls for policy responses to bring the correct incentives for efficient water allocation. These interventions can take different forms, such as water pricing to reflect water scarcity, water trading rules, or a more complex regime of water allocation rules. It has been argued that sometimes local collective action among the commons can lead to efficient monitoring and resource allocations (Ostrom et al., 2003). In the case of floods, the common good characteristics of flood mitigation activities may also lead to overexposure to flood risks and inefficient allocation of risks among exposed individuals and sectors.

Because in a given watershed, water availability is a random variable, policy responses should also be defined contingently not only on delivery time and location but also on "states of nature" such as weather conditions (mostly water availability). This means that in practice water rights should be time-dependent (including spot and future markets) and space-dependent (with water markets delimited within some specific area, e.g. the river basin). Moreover, water rights should be flexible enough to "mimic" state-contingent markets. Importantly, such state-contingent water rights should include the full range of likely possible states of nature, from small departures to average weather conditions to extreme circumstances such as severe droughts and floods. Such consideration of the full range of states of nature should concern preventive actions, *ex ante* (e.g. when building a dam) and *ex post* management actions (e.g. water restrictions, water markets) as part of a comprehensive approach.

2.3. Droughts and floods: Catastrophic risks and incomplete insurance markets

Risk layering

Catastrophic risks are quantitatively and qualitatively different from "normal" risks. Catastrophic risks are generally defined by small probabilities and large damage, often affecting large geographical areas (spatially correlated or systemic risk). These characteristics are often related to difficult insurability and incomplete insurance markets, providing a rationale for policy action in the case of rare and large shocks (whether financial, health or climate) that affect individual farm household income in a way that is beyond the farmer's capacity to cope (OECD, 2009; OECD, 2011). The boundaries between catastrophic and non-catastrophic risks are not well defined but are related to the insurability of the risks and the capacity to cope in the absence of policy measures.

There are several reasons why insurance markets may be incomplete for low probability, high damage and systemic risks. On the demand side, theoretical results show that a risk-averse individual is willing to insure both low and high probability events (Eeckhoudt and Gollier, 1999). However, it is highly debatable that expected utility is the most appropriate framework to capture risk aversion behaviour for low probability high losses events. Experimental evidence tends to show that low probabilities are distorted downward and subject to framing effects by most individuals (Kunreuther et al., 2013). Other research areas such as prospect theory (Barberis, 2013) or probability neglect in behavioural economics point in the same direction.

On the supply side there are several reasons that may lead to low insurance supply for low probability events: the high costs of reinsuring systemic risks, the need to cover adverse selection and moral hazard, and the lack of good information about the likelihood and scope of these rare events. For all these reasons, extreme weather events and catastrophic risks are difficult to insure (Gollier, 1997) and private agriculture insurance markets rarely exist without significant government support (OECD 2011; Mahul and Stutley, 2010). Beyond "normal" and "insurable" risks, there is a category of "catastrophic" risks with which individuals and markets cope with great difficulty. Risk layering (Figure 2.1) can help develop an efficient risk management policy (OECD, 2009; OECD, 2011).

Finally, on both demand and supply sides of the market, climate change tends to increase the uncertainty of risk assessment, since the non-stationary nature of a changing climate impedes using past data to assess future risks. This also affects the boundaries between the different layers of risks that become "moving targets". Such increasing uncertainty is costly for economic agents, who are not only averse to risks, but also to uncertainty. If tools such as insurance and financial markets have been developed with success to allocate risks in an efficient manner, at least to a certain extent, it is a more complex task when deep uncertainty is present.

Catastrophic risks also generate social, media and political pressures to take action. This social demand for "solidarity" after a "catastrophic event" is part of the policy equation. Changes in water allocation regimes and policies during extreme water events require some margin of action. Contingency plans should define the procedures, responsibilities and limits of the policy response (OECD, 2009; OECD, 2011).

For all these reasons, private insurance markets against catastrophic risks are generally not available except in countries with subsidised insurance systems. Indeed, in the face of missing markets, the policy response has often been to provide insurance subsidies. Although this has allowed for the development of crop insurance in several OECD countries, it does not necessarily improve risk allocation, which is the primary motive for any well-designed insurance system. Rather, it redistributes risks between farmers, or between farmers and taxpayers. In this sense, there is a rationale for government intervention to overcome market failure if and only if such intervention provides net benefits compared to a status quo of missing markets (Dixit, 1987; Dixit 1989). Innovations which aim to reduce the implementation cost of drought insurance, such as weather index insurance, exists in several countries (Canada, Spain, United States, and India) but has rarely resulted in a large uptake by farmers[4] – arguably due to basis risk, i.e. the fact that indemnities are not tailored to individual farmer crop losses, but on an imperfect proxy that is the weather index (OECD, 2009; OECD, 2011).

Figure 2.1. The holistic approach to risk layering

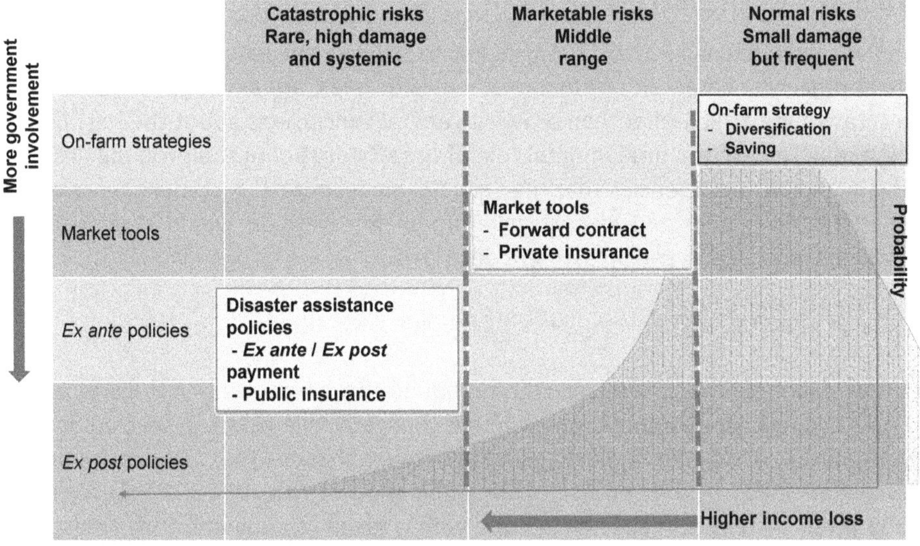

Source: OECD (2011), http://dx.doi.org/10.1787/9789264116146-en.

Dilemmas between ex ante and ex post efficiency

Designing policy responses to incomplete insurance markets for catastrophic risks is a challenging task. In particular, a critical choice for a policy maker when defining policy responses to catastrophic risks is between trying to equalise the expected well-being (the *ex ante* approach) or the expected equalisation of actual well-being (the *ex post* approach). Existing literature recognises that the ranking of the policy options will differ depending on the criterion chosen. More precisely, the two following results are known:

- Every social welfare function[5] is susceptible to *ex ante/ex post* divergence.
- For every such function, there will be choices that are evaluated differently depending on whether the function is applied *ex ante* or *ex post*.

A corollary is that some policies might be optimal from an *ex ante* point of view but not from an *ex post* perspective, and *vice versa*.

Jaffee and Russel (2013) analyse the circumstances in which governments should focus on providing citizens with *ex ante* opportunities to insure against catastrophic risks, or should provide for *ex post* disaster relief policies. They show that *ex ante* and *ex post* optimality are equivalent only when subjective probabilities of risks are identical across all concerned individuals. The intuition for this result is that with divergent beliefs about probabilities for catastrophic events, insurance purchases by individuals will generally differ: the optimists will tend not to buy insurance, and thus be underinsured when an actual event occurs. For them, a system of lump sum relief may be welfare enhancing *ex post*. In such circumstances, Jaffee and Russel (2013) show that by adopting the *ex post* welfare criterion, governments should provide post-disaster aid programmes while prohibiting, or at least containing, *ex ante* catastrophe insurance programs.

The fact that individuals have different subjective probabilities is also supported by empirical evidence. For instance, risk perceptions in the Netherlands concerning the probability of floods differ across individuals in line with different subjective utilities (Botzen et al., 2009). Several explanations for such divergence are available. According to Kunreuther et al. (2013), individuals may not seek information on the probabilities of low-frequency events due to search or transaction costs. Another explanation relates to the high spread of catastrophe bonds, making it difficult for agents to infer the underlying probabilities just from their prices.

The analysis by Jaffee and Russel (2013) seems to strongly challenge the commonly accepted view that *ex ante* approaches, based on risk management principles, are always preferable to *ex post* relief following a catastrophic risk. Rather than drawing general conclusions about the best policy responses, it sheds interesting light on the fundamental role of uncertainty about the probabilities of catastrophic risks and the economic inefficiencies that may arise in such context. A comprehensive analysis of the costs and benefits of *ex ante* and *ex post* government intervention would require comparing their effects on farmers' production and investment decisions.

2.4. Market failures, vulnerability and resilience to droughts and floods

Notions of vulnerability and resilience have gradually become central in the area of policy analysis of climate risks and, perhaps even more so, climate change. While risks can be traditionally defined as a combination of probability and impacts, vulnerability emphasises the "propensity or predisposition to be adversely affected" by a given risk or a combination of risks (IPCC, 2012). Vulnerability is the property of a given system: its capacity to cope with a given (combination of) risk(s). Weather risks such as droughts and floods are the joint outcome of weather events, exposure to these events, and vulnerability of affected systems. Risk vulnerability is typically considered to be the joint outcome of risk exposure and risk sensitivity of the system considered (OECD, 2014a).

Vulnerability is a relative concept, which makes it difficult to assess or quantify (FAO and OECD, 2012). One can make difference between physical vulnerability, which refers to the sensitivity of cropping and livestock systems to droughts and floods and can be measured in terms of production losses, and economic vulnerability, which refers to the capacity of the farm system to absorb income losses due to these extreme events. For example, a farm system may be physically vulnerable to droughts while at the same time have a reasonable capacity to absorb the associated income shock through well-designed insurance systems, or the use of precautionary saving financial tools.

A first-order issue in analysing vulnerability is to have a clear definition of the system considered. Droughts and floods typically affect several economic sectors and people, such as urban areas, agriculture, industry, and ecosystems. They can also have indirect economic impacts outside the affected areas by affecting demand and supply on related markets: examples include increases in agricultural commodity prices due to drought induced crop failure in a large production area, and the need to import feed for livestock. Physical and economic interdependencies associated with specific characteristics of water imply there can be synergies and trade-offs in vulnerability reductions across water users and uses. For example: reducing flood risk for urban areas through hydrological infrastructures can increase flood risks on agricultural land areas; water restriction rules can reduce vulnerability of certain water users to drought risk, but at the cost of increasing vulnerability for other users and uses; and improvements of water use efficiency in a given sector can potentially reduce drought risk vulnerability for all water users, by making more water available for these users (Figure 2.2).

Vulnerability of hydrological and agricultural systems to droughts and floods depend on policy responses that are in place to mitigate the impacts of these risks on the concerned sectors, and the actions of affected agents. Beyond sectoral policies, policy responses to market failures related to water management and insurance markets also have an important role to play as they affect the interdependencies of the system. Water allocation mechanisms (e.g. water rights, water pricing, collective arrangements) can be designed to ensure that water is allocated in an economically efficient way, taking into account environmental externalities so that water is delivered in priority to uses with the highest social values. Finally, costs of drought and flood risks, when possible, can be mutualised geographically through insurance or compensation mechanisms or smoothed in time through dedicated precautionary funds or saving policies that target these risks. Taken together, these policy responses affect the overall level and distribution of the costs of droughts and floods to the concerned individuals and economic sectors, and their associated exposures and vulnerabilities.

The notion of resilience is related to vulnerability, even though it focuses on the time dimension. A system is considered as resilient if it is able to recover following a shock. The more a system is vulnerable, the less in principle it will be able to recover. However, resilience also deals with the long term ability of the system to cope with risks. Notably, if drought and flood risks are increasingly frequent, this may progressively erode the capacity of the agricultural sector to recover, and hence would threaten its long-term resilience. This is especially the case for droughts, which can last longer than a single cropping season.

There can be important trade-offs between vulnerability in the short-run and vulnerability in the long term. In the context of droughts and floods, a typical example is the reliance of irrigators on groundwater resources, which can provide a form of self-insurance against surface water shortage in the short-term, but can *in fine* lead to overexploitation of the resource in the longer term. Another example are crop insurance subsidies against weather risks: while artificially reducing the cost of risks borne by farmers, they can ultimately lead to greater risk exposure by making them invest in more risk-prone farming practices and, in the longer term, make farming systems more vulnerable to risks as compared to a situation with no insurance subsidies.

Figure 2.2. Schematic representation of a generic drought onset and associated socio-economic direct costs

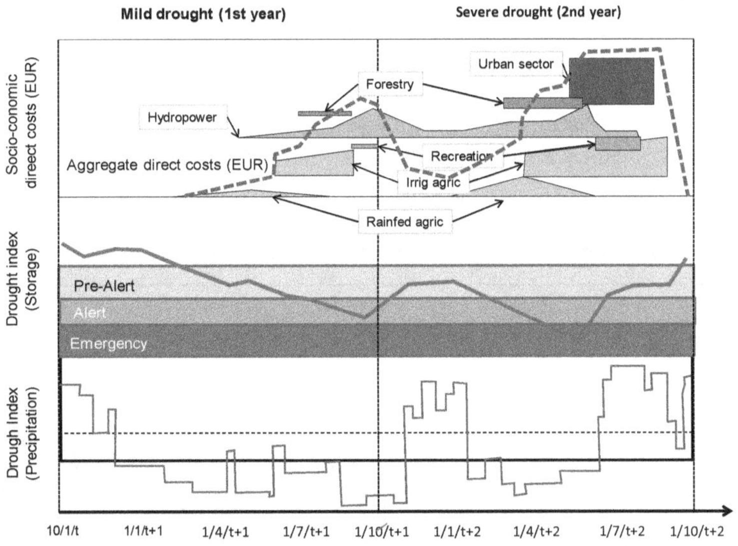

Source: Garrido et al. (2012).

2.5. Policy and market drivers affecting vulnerability of agriculture to droughts and floods

Droughts and floods policies in agriculture take place in a broader policy, market and environmental (climatic notably) context. Farmers' production and investment decisions, and the associated risk exposure and vulnerability that arise, are driven by these policy, market and environmental drivers. Two projected trends are in particular likely to exacerbate the importance of drought and flood risks for the agricultural sector, and could require specific and adapted policy responses. The first trend is climate change, which is expected to increase the frequency and intensity of extreme events and the second one is the increasing competition between agriculture and other sectors for two natural resources: water and land. Major drivers include the potential rise in commodity prices and agricultural policies supporting irrigated crops with high water consumption per unit of agricultural land, or the increasing competition for land use across agricultural, forest and urban areas. Analysis of these major market and policy drivers is beyond the scope of this study; however, they suggest that policy approaches to managing droughts and floods are likely to take on greater importance in the context depicted by projections.

Climate change

The issue of extreme weather events is gaining the attention of policy makers in all economic sectors (IPCC, 2012; OECD, 2011). Although there are several ways to define and measure these events (Sheffield et al., 2012), it is generally expected that climate change will increase their frequency and intensity (IPCC, 2012; OECD, 2014b). Indeed, recent trends in the last decades seem to indicate an increase in the frequency and severity of extreme weather events, especially in the cases of floods and droughts (Figure 2.3).

According to the World Meteorological Organization (World Meteorological Organization, 2013), 2001-2010 has been the warmest since 1850 in terms of average temperature, and nearly half of the maximum temperatures recorded since 1961 have been in the last decade. In the early 2000s, there were many droughts in western United States, southeast Australia, and the People's Republic of China. More recently, severe droughts in the Russian Federation and central United States have occurred. However, overall scientific evidence about past trends in droughts and their characteristics are less clear. Sheffield et al. (2012) and Spinoni et al. (2013) show a small increase in frequency and duration of droughts at the global level, but contrasting projections in different regions of the world. Some countries are experiencing increasing frequency and severity of droughts, while others have witnessed a decreasing trend of these events.

Figure 2.3. Global weather-related disasters, 1980-2009

Number of disasters

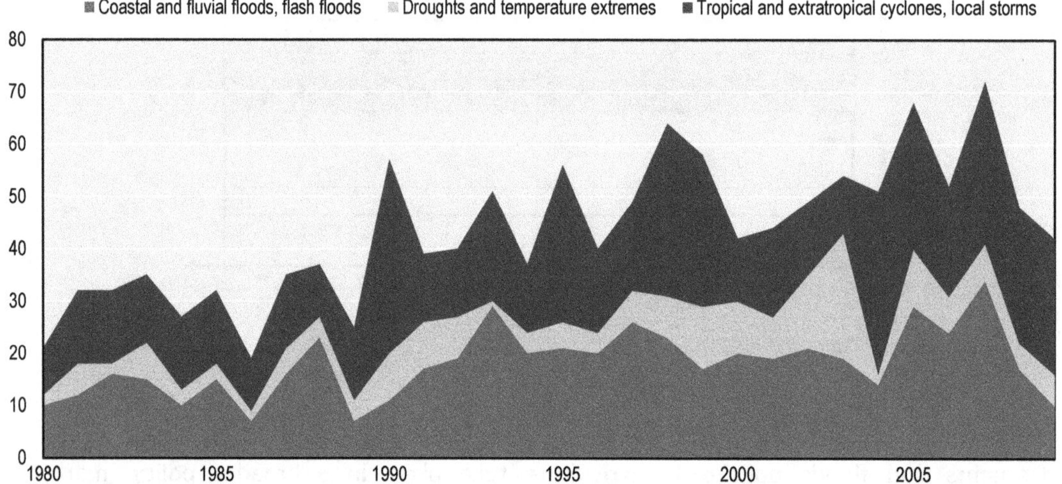

OECD (2012), Global weather-related disasters, 1980-2009, in *OECD Environmental Outlook*, http://dx.doi.org/10.1787/env_outlook-2012-graph71-en.

There are few global assessments of financial damage due to floods. The Dartmouth Flood Observatory lists 857 flood events in OECD countries for the period 1985-2008. The average damage caused by extreme rainfall events was USD 1.1 billion per event (based on 191 events for which damages were estimated), thus it is possible that extreme rainfall alone has been responsible for USD 700 billion of damages over the last 13 years (Morris et al., 2010). The *OECD Environmental Outlook* (OECD, 2012) estimates that the number of people and value of assets at risk from floods will be significantly higher in 2050 than today.

In addition to the other factors expected to modify their economic environment (e.g. increasing population, resource scarcity), farmers and water risk management systems will operate in a *non-stationary climate*. This poses serious challenges for adaptation, and in particular for agricultural water management on both rain-fed and irrigated lands.

Increasing competition for water and land resources

Agriculture represents on average 44% of total freshwater withdrawals and 36% of total land area in the OECD zone (OECD, 2013), playing a major role in the management of water systems and in land use in most OECD countries. The interdependence of agriculture and other economic sectors is thus particularly worth considering for the management of water systems in times of extreme events. Farm cropping systems can exacerbate the consequences of floods and droughts, but also contribute to reducing the harmful consequences of floods in urban areas by providing water retention services and slowing water flows across farmland. In situations of water stress, the different water users must solve collectively the problem of allocation with an appropriate set of sharing rules, regulatory and economic instruments, and collective arrangements. There is a need to manage these interdependencies on an appropriate scale. These interdependencies are likely to become more acute with economic development, demographic growth, and climate change, and the projected rise in frequency and intensity of extreme weather events.

Notes

1. In economics, a positive risk premium corresponds to risk averse preferences. Risk averse preferences are related to the concavity of the utility function, as economic agents do not value equivalently an additional unit of income depending on their initial levels of wealth.

2. Several risk-reducing techniques, such as crop diversification of drought-resistant seeds, do not necessarily require farmers to be risk averse in order to be adopted. They can be adopted by risk-neutral farmers if their expected benefits are higher than their expected costs. However, risk aversion may be an additional incentive for such adoption.

3. Rivalry is particularly relevant for consumptive uses, such as irrigation water withdrawals, but can also concern other types of water uses as far as water quantity is concerned; for example, minimum environmental flows. This is less clear in the case of recreational activities, where water quantity (the flow of a river or the level of a lake) may matter, but the different types of activities relying on water support — swimming and fishing for example — are not necessarily themselves in rivalry. Rivalry can also affect non-consumptive use of water is such uses occur simultaneously between the concerned water users. Excludability may depend on the organisational level, the physical characteristics of the watershed, etc., and could be enforced outside a given community and not by its members. These variants in the physical and socio-economic circumstances may explain or even justify on the grounds of efficiency the existence of different institutional organisations and policy tools.

4. The principle of index-based insurance is to base the calculation of indemnities on the realisation of a given climatic variable, such as rainfall or temperature – the weather index – rather than on the insurees' real losses. It is grounded in the fact that there is a correlation between the weather index and insurees' losses. This is especially true in the case of systemic risks, i.e. risks that are highly spatially correlated in a given area, which is often the case of droughts. The main advantage is to reduce administrative costs linked to the monitoring of insurees' losses, as well as moral hazard costs. However, the imperfect correlation between the weather index and insurees' losses means the indemnification may not always correspond to the real losses experienced by farmers. Such residual risk borne by the farmer is called basis risk.

5. The social welfare function maps each utility vector onto a number, which represents the position of that outcome in the ordering.

References

Barberis, N. (2013), “Thirty Years of Prospect Theory in Economics: A Review and Assessment”, *Journal of Economic Perspectives,* Vol. 27, pp. 173-196.

Botzen, W.J.W., J.C.J.H. Aerts and J.C.J.M. van den Bergh (2009) “Dependence of flood risk perceptions on socioeconomic and objective risk factors”, *Water Resources Research*, Vol. 45(10), pp. 1–15.

Dixit, A.K, (1989), "Trade and Insurance with Adverse Selection," *Review of Economic Studies*, Wiley Blackwell, Vol. 56(2), pp. 235-47.

Dixit, A.K. (1987), "Trade and insurance with moral hazard," *Journal of International Economics*, Elsevier, Vol. 23(3-4), pp. 201-220.

Ehrlich I. and G. Becker (1972), “Market insurance, self-insurance and self-protection”, *Journal of Political Economy,* Vol. 80, pp. 623–648.

Eeckhoudt L. and C. Gollier (1999), “The Insurance of Low Probability Events”, *Journal of Risk and Insurance*, No. 66, pp.17-28.

FAO and OECD (2012), *Building resilience for adaptation to climate change in the agriculture sector: Proceedings of a Joint FAO/OECD Workshop*, FAO and OECD, www.oecd.org/tad/sustainable-agriculture/climate-change-resilience-agriculture-workshop.htm.

Froot, K., D. Scharfstein, and J. Stein (1993), “Risk management: Coordinating corporate investment and financing policies”, *Journal of Finance*, Vol. 48, pp. 1629-1658.

Garrido A., N. Hernández-Mora and M. Gil (2012), “Drought impact assessments”, unpublished manuscript.

Gollier, Christian (1997), “About the Insurability of Catastrophic Risks”, *The Geneva Papers on Risk and Insurance*, Vol. 22, pp. 177-186.

IPCC (2012), *Managing the Risks of Extreme Events and Disasters to Advance Climate Change Adaptation. A Special Report of Working Groups I and II of the Intergovernmental Panel on Climate Change*, Cambridge University Press, Cambridge, and New York.

Jaffee, D. and T. Russell (2013), "The Welfare Economics of Catastrophe Losses and Insurance," in *The Geneva Papers on Risk and Insurance - Issues and Practice*, Palgrave Macmillan, Vol. 38(3), pp. 469-494.

Kunreuther, H.C., M.V. Pauly, and S. McMorrow (2013), *Insurance and Behavioral Economics: Improving Decisions in the Most Misunderstood Industry*, Cambridge University Press, Cambridge.

Mahul, O. and C.J. Stutley (2010), “Government Support to Agricultural Insurance”, The World Bank, Washington, DC., https://openknowledge.worldbank.org/bitstream/handle/10986/2432/538810PUB0Gove101Official0Use0Only1.pdf?sequence=1.

Morris, J., T. Hess and H. Posthumus (2010), "Agriculture's Role in Flood Adaptation and Mitigation: Policy Issues and Approaches", in OECD, *Sustainable Management of Water Resources in Agriculture*, OECD Publishing, Paris. http://dx.doi.org/10.1787/9789264083578-9-en.

OECD (2014a), *Disaster Risk Assessment and Risk Financing*, OECD Publishing, Paris, http://www.oecd.org/gov/risk/G20disasterriskmanagement.pdf.

OECD (2014b), *Climate Change, Water and Agriculture: Towards Resilient Systems*, OECD Studies on Water, OECD Publishing, Paris, http://dx.doi.org/10.1787/9789264209138-en.

OECD (2013), *OECD Compendium of Agri-environmental Indicators*, OECD Publishing, Paris. http://dx.doi.org/10.1787/9789264186217-en.

OECD (2012), *OECD Environmental Outlook to 2050: The Consequences of Inaction*, OECD Publishing, Paris, http://dx.doi.org/10.1787/9789264122246-en.

OECD (2011), *Managing Risk in Agriculture: Policy Assessment and Design*, OECD Publishing, Paris. http://dx.doi.org/10.1787/9789264116146-en.

OECD (2010), *Sustainable Management of Water Resources in Agriculture*, OECD Studies on Water, OECD Publishing, Paris. http://dx.doi.org/10.1787/9789264083578-en.

OECD (2009), *Managing Risk in Agriculture: A Holistic Approach*, OECD Publishing, Paris, http://dx.doi.org/10.1787/9789264075313-en.

Ostrom, E. (2005), *Understanding Institutional Diversity*, Princeton University Press, Princeton, http://press.princeton.edu/chapters/s8085.pdf.

Ostrom, E., P. Stern and T. Dietz (2003), “Water rights in the commons”, *Water Resources Impacts*, Vol. 5 (2), pp. 9-12.

Sheffield, J., K.M. Andreadis, E.F. Wood, and D.P. Lettenmaier (2009), “Global and continental drought in the second half of the 20th century: severity-area-duration analysis and temporal variability of large-scale events”, *Journal of Climate*, Vol. 22(8), pp. 1962-1981.

Spinoni, J., G. Naumann, H. Carrao, P. Barbosa, J. Vogt (2013), “World drought frequency, duration, and severity for 1951–2010”, *International Journal of Climatology*, Vol. 34, pp. 2792–2804.

Starrett, D.A. (2003), Property Rights, Public Goods and the Environment”, Chapter 3 in Mäler, K.G. and J.R. Vincent, *Handbook of Environmental Economics*, Vol. 1: Environmental Degradation and Institutional Responses, Amsterdam.

World Meteorological Organization (2013), *The Global Climate 2001-2010: A Decade of Climate Extremes – Summary Report*, WMO-No. 1119, World Meteorological Organisation.

Chapter 3

Policy approaches for the sustainable management of droughts and floods in agriculture

This chapter reviews and analyses policy approaches that foster efficient, resilient and sustainable management of droughts and floods in agriculture. These include water risk mitigation policies in the short- and long-term. It also reviews compensation and insurance policies, and provides elements of comparison between policy approaches of five OECD countries: Australia, Canada, France, Spain and the United Kingdom.

A risk management strategy against droughts and floods should include a comprehensive set of actions that can be taken *ex ante* or *ex post,* at either individual or collective level. Farmers can individually reduce their risk exposure and vulnerability to droughts and floods by modifying their cropping systems, adopting drought-resistant seeds, or appropriately managing floodplain areas. When available, farmers can also rely on risk sharing tools such as agricultural insurance against droughts or floods.

On-farm, individual water risk management strategies, although essential, are insufficient to manage drought and flood risks in an efficient way. As explained in Chapter 2, water-related risks have specific features which are likely to impede efficient risk prevention and allocation because of a set of market, government and behavioural failures that require specific, targeted policy responses. Public policies in the area of water management are widely recognised as indispensable to complement individual mitigation activities, or to orientate them in the sense of social welfare and sustainability.

To facilitate the analysis of the wide range of policy approaches addressing flood and drought conditions in agriculture from a holistic perspective, the following categories of policies are analysed. Such categorisation should not be seen as set in stone or generalised to other contexts or purposes; and the distinction between categories may in some case be subject to discussion. However, it provides a simple entry point for reviewing policy tools that are, or could be used, and their eventual complementarity or substitutability:

- ***Water risk mitigation policies*** which can be divided into:
  - *Long-term water risk mitigation policies*, which aim to mitigate droughts and floods through *ex ante* preventive actions that target the fundamental drivers of risk exposure and vulnerability, such as water demand and supply policies, land regulation, and water planning.
  - *Short-term water risk management policies*, which target the management of droughts and floods events in the short-term. They can be defined as the set of policy responses undertaken to mitigate the consequences of extreme water events once these events have occurred. Examples include water restriction rules during a drought, or emergency plans for floods.
- ***Compensation and insurance policies***, which focus on mitigating income losses incurred by farmers following drought and flood events through financial compensation.

3.1. Water risk mitigation policies: Droughts

Reducing structural water stress through supply and demand water policies

Drought risk vulnerability is often the outcome of a structural imbalance between supply and water demand in a given water basin. Overuse and misallocation of water resources due to incomplete or over-allocated water rights, under-pricing of water and deficiencies in long-term resource planning contribute to this imbalance, which in turn mechanically translates into higher water shortage risks. In water basins where there is structural water stress, water shortages are more likely to arise from even mild precipitation deficit. In short, water stress generates water shortage risks, and thus vulnerability to droughts.

Water policies that aim to reduce structural water stress, even when they do not specifically target drought events but focus on average climate conditions, are essential to reducing the frequency and severity of water shortage risks. Market failures in water resource allocation tends to lead to overuse of water, and thus mechanically to excessive risks of water shortage. Water policies are typically divided into demand-based policies, such as water rights and water pricing, aimed at a better adjustment of water uses with water availability, and supply-based policies, such as water storage infrastructures (dams, reservoirs) which can play a complementary role by increasing the average level of water resource for users as well as regulate water flows. Up to the early 1990s, increasing water supply by developing hydrological infrastructures used to dominate the water policy landscape. Since then, the paradigm has been to shift towards demand-based water policies, which are generally considered as more cost-efficient and sustainable in the long-term (OECD, 2010a).

The combined role of demand and supply water policies "at the mean" on the risk of water shortage is illustrated schematically in Figure 3.1, which represents a hypothetical probability distribution (density) of water supply together with a level of water demand which is assumed to be non-stochastic. At date 1, the levels of water supply and demand are represented by, respectively, S1 and D1. Hence, the probability that demand falls beyond supply is equal to *a+b+c+d*. Suppose that at period 2 demand decreases to a level described by line D2, while the probability distribution shifts on the right, from S1 to S2. In this case, the probability that demand falls beyond supply is reduced to *c*. In the case where only demand shifted from D1 to D2, while keeping supply at S1, this probability would have been higher and equal to *a+c*. One can expect climate change to reduce the average and increase the riskiness of water supply, while at the same time water demand is projected to increase.

OECD countries are currently characterised by various degrees of water stress at national levels, with half of these countries facing moderate to medium-high water levels (Figure 3.2). Water stress is calculated as the ratio between gross water abstraction and total renewable resources (OECD, 2013). Water stress is relatively higher than the OECD average in southern European countries such as Italy, Spain, Portugal, and Greece, but also in western European countries such as Belgium, France and Germany. Water stress is also significant in Japan and Turkey. These national figures do not account for local and seasonal variations, which can be of first order of importance.

Figure 3.1. Water demand and supply and probability of water stress

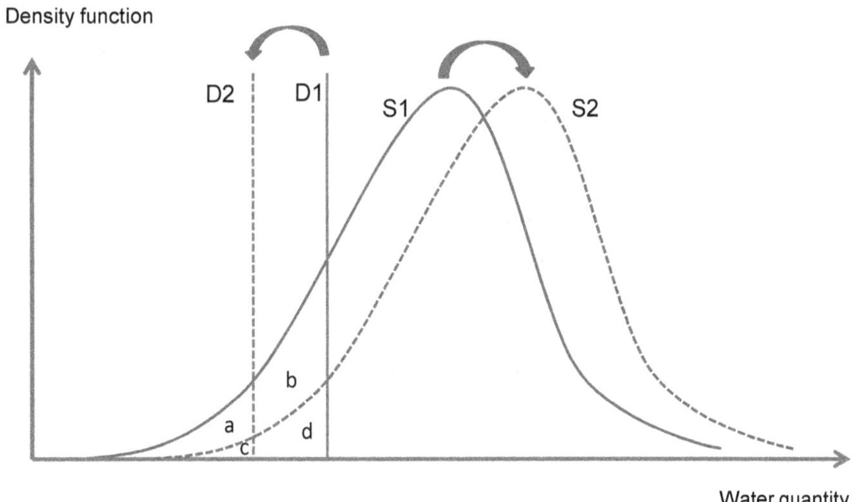

Figure 3.2. Intensity of use of freshwater resources

Gross abstractions as percentage of total renewable resources

Water stress: <10: low; 10%-40%: moderate and medium-high; >40%: high

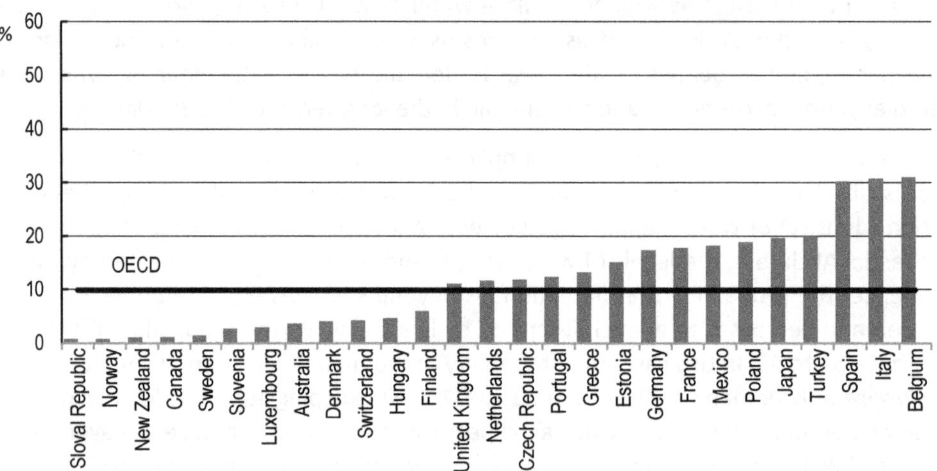

Note: The black horizontal line represents OECD average.

Source: OECD (2013), http://dx.doi.org/10.1787/9789264185715-en.

In several OECD countries, agriculture represents the largest share of total freshwater withdrawals, due especially to irrigation, thus playing a major role in the overall level of water stress. With an average 44% of total freshwater withdrawals in the OECD area, the share of agriculture can be even higher in certain countries, especially during the summer seasonal peak when agriculture can account for more than 80-90% of total freshwater withdrawals (Hoekstra et al., 2012).

As regards trends, it is noticeable that for most OECD countries, freshwater withdrawals have tended to decrease or remain stable over the last 15-20 years. Agricultural water withdrawals have followed similar tendencies and as a consequence, average water stress has tended to be stable (Spain, France, and Mexico) or even to have decreased, e.g. Israel (OECD, 2013). The reduction of agricultural water withdrawals comes from either a contraction of irrigated areas or greater water use efficiency, or a combination of the two. Despite these encouraging trends, projections for the next decades underline the risks of increasing water stress in many water basins of the OECD area, due to a combination of rising water demand and decreasing water availability linked to climate change.

Water demand policies are generally based on a system of water entitlements, either individual or collective on the basis of which economic instruments can be developed, e.g. water pricing and tradable water rights. Administrative water pricing is typically implemented by a government or collective agency that provides a water service and attempts to recover at least part of the costs. In practice, water entitlement systems can be applied to different categories of users and based on priority of use (e.g. seniority) with unconstrained consumption, or based on quantities or volumes of water defined individually or collectively, daily or weekly flows, etc. Such systems are widespread in OECD countries, but their characteristics vary a great deal, even between regions and irrigated perimeters within the same country. In particular, formal water quota trading is still limited worldwide including in OECD countries, although there is a trend towards the development of such markets or refining existing ones (Australia, United States). Nevertheless, a recent OECD survey on water allocation systems shows that a significant number of allocation regimes allow for trading, leasing or transferring water in one way or another, although a variety of conditions are applied (OECD, 2015a).

Different types of policy instruments have been used in the past 20 years to regulate water demand in agriculture (OECD, 2010a). Although some types of instruments dominate in some countries, in most cases water demand policies in agriculture combine a set of regulatory tools, flexible mechanisms, either formal or informal, and economic incentives. For example, Australian water markets are substantially regulated to ensure their proper functioning, with rules governing trading possibilities among Australian states, government buy-backs of water rights, or the definition of minimum environmental flows. Conversely, in some countries where regulatory, command-and-control approaches seem predominant, it is not unusual to observe in practice flexible implementation in the allocation of water rights in times of shortage through, for example, a negotiation process between water users during a drought event.

Despite undeniable progress in several countries, there is still significant room for improvement (Box 3.1). In most OECD countries, water cost recovery in agriculture is incomplete concerning operation, maintenance, and investment costs. This raises issues of financial sustainability and unstainable incentives to water overconsumption given that, at their current level, water cost recovery do not include the cost of negative externalities (OECD, 2010a). Full recovery of costs related to water use is notably at the core of EU policies in this area since the publication of the *Water Framework Directive*.

Rising concerns in the European Union about water scarcity and droughts led the European Commission (EC) to propose in 2007 a set of policy responses to these issues in the form of a Communication paper to the European Parliament and the Council (European Commission, 2007a; 2007b; 2012; Box 3.2). The EC Communication underlines the need to reinforce the "quantity" dimension of water management to improve the mitigation of water scarcity and drought vulnerability in Member States. Although specific drought management tools were proposed, the Communication reaffirmed that policy measures targeting structural water stress, such as putting the right price tag on water, allocating water funds more efficiently, and financing water efficiency, constitute priority measures of an efficient and comprehensive policy mix to mitigate drought risks and water shortage in the European Union.

Box 3.1. Recent trends in water policies in OECD countries

Most OECD countries have begun to reform water policies in the last decade, although progress has varied. The main trends observed in the OECD area can be summarised as follows (for more detail, see OECD, 2010a).

- Trend towards decentralisation of water management policies.
- Shift from policies focused on water supply towards to demand-based policies.
- Development of economic tools such as water pricing and water markets in certain countries, mostly within the agricultural sector. Inter-sectoral transfer of water remains infrequent.

Despite these general trends, water pricing remains limited or partial in OECD countries, and where they exist cover mostly operations and maintenance costs. Scarcity costs are seldom included in a water pricing formula. Some countries have chosen to develop water quota trading, notably the United States and Australia. The reforms undertaken in some countries have contributed (agricultural policies and markets notably) to an increase in water use efficiency in the agricultural sector. For instance, irrigation water application rates, which measure the quantity of irrigation freshwater withdrawals per unit area of irrigated land, have decreased sharply in several OECD countries, such as Australia and Mexico.

This progress could, however, be challenged in the next decade by the rise in demand for agricultural products in the context of a growing world population, increased incomes, and climate change.

Source: OECD (2010a), http://dx.doi.org/10.1787/9789264083578-en.

Box 3.2. The 2007 initiative of the European Union to manage drought risks

In a context of rising episodes of drought and associated water shortages, the European Commission proposed seven policy options in a Communication *Addressing the challenge of water scarcity and droughts in the European Union* (European Commission, 2007c):

1. Putting the right price tag on water
2. Allocating water and water-related funding more efficiently
3. Improving drought risk management
4. Considering additional water supply infrastructures
5. Fostering water efficient technologies and practices
6. Fostering the emergence of a water-saving culture in Europe
7. Improve knowledge and data collection.

There is a follow-up process of this Communication in the form of annual reports.

Since the 2007 EC Communication, follow-up reports of the progress made by Member Countries have been produced about every year, accompanied by a study analysing water policy gaps in the European Union. According to these sources, although progress has been made in certain Member countries in the different areas of policy responses to drought and water scarcity, there remain significant policy gaps. In addition, it is difficult to measure the real impact of policy reforms, e.g. expenses on water use efficiency due to the lack of data and indicators.

In several countries, a particular issue for water stress is the heavy reliance of farms on groundwater. As underlined in several OECD reports (OECD, 2010a; OECD, 2015b), there is a lack of regulation of groundwater resources in many OECD countries as compared to surface water resources. Although the reasons for this are complex, the consequences on the sustainability of groundwater resources could be negative. This is particularly the case in circumstances where there is a shortage of surface or snowpack water that results in the use of groundwater as a water supply substitute, acting as an implicit form of "water insurance" for farmers. The 2013-14 droughts in California are an illustration of this: groundwater has been used as a substitute to insufficient surface water to such levels that it may negatively affect in the future the ability of groundwater systems to recharge as quickly as before (Howitt et al., 2014; OECD, 2015b).

Water use efficiency: necessary but not sufficient

In view of a successful strategy to manage the water scarcity challenge, there has been a strong emphasis in the last two decades in policy and academic circles on water use efficiency as a means to reduce structural water stress and vulnerability to the risk of water shortage (FAO, 2008). Water use efficiency has different components, which are summarised in Figure 3.3. Different technical options are potentially available to improve water use efficiency (Pereira et al., 2012; Levido et al., 2014; Jensen et al., 2014) which would allow the agricultural sector to produce more food, feed and fibre while also freeing up water resources for other users and uses. The discussion is closely linked to the debate on water productivity gaps, i.e. the fact that current irrigation practices are far from optimal and that moving to more efficient irrigation practices (e.g. from gravity to sprinkler) would allow for substantial water savings (Brauman et al., 2013; Gómez and Pérez-Blanco, 2014; Scheierling et al., 2015) .

Figure 3.3. Processes influencing irrigation efficiency

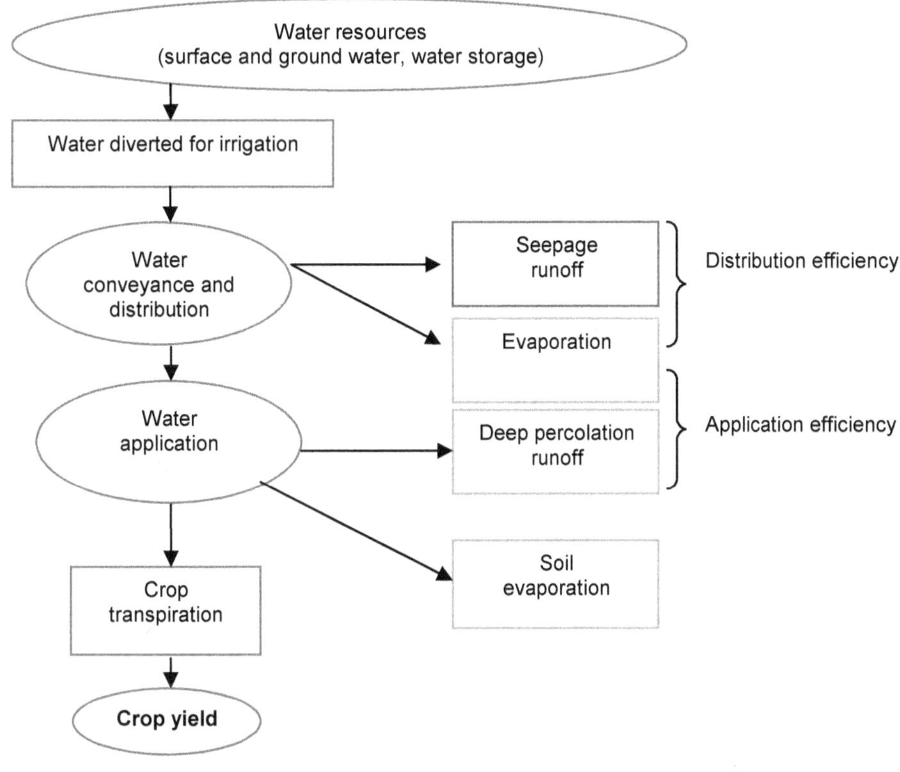

Source: Based on Pereira et al. (2012), http://econpapers.repec.org/article/eeeagiwat/v_3a108_3ay_3a2012_3ai_3ac_3ap_3a39-51.htm.

Improving water use efficiency is a policy objective in many OECD countries facing situations of water stress and vulnerability to droughts. It is notably the case of the European Union for which the objective of fostering water efficient technologies and practices is among the proposed policy options to better manage water scarcity and droughts (Box 3.2). In the United States, the Environmental Quality Incentives Program (EQIP) is a significant source of federal financing for the modernisation of irrigation practices and the improvement of water use efficiency. In Australia, water policy reforms of the last decade not only introduced water markets, but also included substantial programmes aimed at modernising irrigation infrastructures (Kirby et al., 2014).

While improving water use efficiency is necessary to move forward a green growth strategy in agriculture, several issues must be addressed to ensure that policy approaches achieve their objectives. A too narrow focus on water use efficiency, together with a lack of water policy coherence could lead to perverse effects and counterproductive outcomes. Three issues are of particular concern in this area (Gómez and Pérez-Blanco, 2014):

- the hydrological paradox
- the risk of rebound effect
- the indirect impact associated with production choices.

The first challenge is the risk of "hydrological paradox". Increasing water efficiency may indeed be neutral, or even reduce water availability for other users. In many watersheds, a large share of irrigation does return to the water system, allowing for groundwater recharge or contributing to downstream river flows. Increasing on-farm water use efficiency is expected to reduce agricultural water withdrawals, but also the associated returns to the water system. This phenomenon is sometimes called the "hydrological paradox" and illustrates the sometimes beneficial role of irrigation to water systems. Mitigating these unintended consequences of water use efficiency gains implies an appropriate water accounting at the basin scale that considers not just withdrawals but also water returning to the system. Moving from hydrological science to the inclusion of such return flows in water right systems is, however, a complex task. Accounting for return flows should thus be studied more systemically to assess their relative importance in watersheds. In a second step, return flows would need to be accounted for in water allocation systems to better reflect overall water supply and demand, and thus improve the efficiency of water allocation (OECD, 2015a).

The second potential unintended consequence is the risk of a rebound effect, also called the Jevons' paradox. It finds its origin in the area of energy economics although research increasingly suggests its potentially important relevance in water management in agriculture. The idea is simple: improving water use efficiency increases the profitability of the associated irrigated crops, and thus incites farmers to expand their irrigated land areas. This happens when water savings arising from efficiency gains are captured by the farmer, rather than returned to the water system. The classical corollary is that water use efficiency gains should be accompanied by a regulation of water demand or irrigation areas to prevent this rebound effect from occurring.

Finally, even taking into account the previous risks of perverse effects, investments to increase water use efficiency can incite farmers to follow a path of specialisation in irrigated crops, which in the end would make them more dependent on water resources and the risks associated with climate change.

The role of water storage infrastructures and alternative sources of water supply

Supply-based approaches tended to dominate up to the late 1980s, after which the emphasis progressively moved to demand-based policies in a context of projected increases in water demand and scarcity, and climate change. Despite this tendency, the increase and securitisation of water resources is still an important dimension of the water management debate. In general, dams modify the statistical distribution of water supply in more complex ways, and often serve several purposes; increasing water supply is one, but they also reduce the risks of floods and droughts at the extremes of the distribution by regulating the fluctuation of water flows. Regulation of water flows or flood prevention can be a significant, and for certain dams its primary purpose. Indeed, the value of water resources can fluctuate highly in time and increasing water supply is often be synonymous with maintaining water flows during the summer period.

Several OECD countries, such as Australia, France and Spain, are considering dams, and more generally water storage infrastructures, as key elements of their water scarcity and drought policies. Australia is particularly interested in developing the integration of water storage management with existing water markets, thus focusing on management issues. Spain plans to develop new hydrological infrastructures to respond to increasing water scarcity. In France, the policy debate has focused on the use of on-farm water storage as a means of securing access to the resource. Israel places particular emphasis on the potential of desalination techniques as an alternative source of water supply. In the European Union, water storage infrastructures can provide water for some uses, but the policy priority is given to measures that aim at regulating agricultural water demand and increasing water use efficiency in a way that is compatible with a well-functioning ecosystem (Box 3.2). Beyond these specific national approaches, issues that are common to many OECD countries are ageing infrastructures and the need to finance modernisation plans.

The policy challenge with water storage infrastructures seems to have evolved in recent years. One the one hand, water storage infrastructures such as "big" dams were criticised in the past because of their negative consequences on ecosystems, their high financial cost and low return to investment (Ansar et al., 2007), and more recently their unintended negative social implications (Duflo and Pande, 2007). These criticisms continue and have become even more important since environmental concerns have become a major issue world-wide. On the other hand, increases in water scarcity and extreme water events associated with the projected acceleration of the water cycle due to climate change may lead certain countries to re-examine the costs and benefits of their water storage policy strategies for adaptation purposes. A recent example is the drought in California, in which the development of alternative sources of water supply is discussed among the set of possible policy responses. Ensuring food security can also play an important in the overall political balance regarding water storage strategies.

One cannot provide too general recommendation on the relevance of developing water storage infrastructures. The best policy recommendation in this area is that water infrastructures projects should involve a transparent and inclusive decision-making process, in which the full set of costs and benefits for different water users and uses (including ecosystems) are clearly recognised using state-of-the-art of cost-benefit analysis (OECD, 2006). Drawing on lessons from previous failures to estimate the real costs of these projects could be useful in that regard. Considering more ecosystem-friendly forms of water storage, such as floodplains, could in certain areas be more cost-effective and sustainable than "grey" infrastructure. The challenge is to ensure that all possible alternatives are considered in the assessment during the initial steps of a project development, and thus avoid any *ex ante* bias against green infrastructure options. Having a more standardised and harmonised methodology to assess the full range of costs and benefits could help overcome this bias in practice (Chapter 1). In agriculture, political economy considerations play a strong role in this area in the sense that developing water infrastructures are sometimes considered as a counterpart to water reforms that impose more stringent regulations on water demand.

Water shortage management plans

Beyond the existing and desirable efforts undertaken to mitigate drought risk through general water policies and specific prevention measures, there is still a need to manage drought events when they occur so as to minimise their associated costs for water users. This short-term management of risk, which takes place *ex post* or during the course of the event, can be as important as *ex ante*, long-run mitigation policies, although it seems to be that they are less studied in economic literature.

As regards terminology, some authors use the term "adaptation" to describe these short-term responses, while others prefer the term "crisis management" (Morris et al., 2010; Lefebvre and Thoyer, 2012). The word "crisis" is often associated with the idea of unpreparedness and suddenness, but what is commonly called "crisis management" refers to a set of pre-established decision rules and the capacity to deal effectively with the situation as it arises. It seems preferable to use the term "risk management plans" to describe pre-defined decision rules to apply in circumstances of droughts and floods, while the term "crisis" is better suited to reflect *ad hoc* management associated with some form of incompleteness of pre-defined rules that is revealed in the course of action.

Agricultural freshwater withdrawals are highly variable during the growing season and typically reach a peak during the crucial phases of crop growth, notably the flowering period. During these peak periods, they can represent up to 80-90% of total freshwater withdrawals, while the average share over the whole year is much lower. Short-term management of water shortage deals with these limited but critical time periods when water shortage reaches a peak, and the consequences for agricultural production as well as other water users and uses can be significant.

Figure 3.4. Crisis management: Key steps in the management of temporary water shortages and excesses

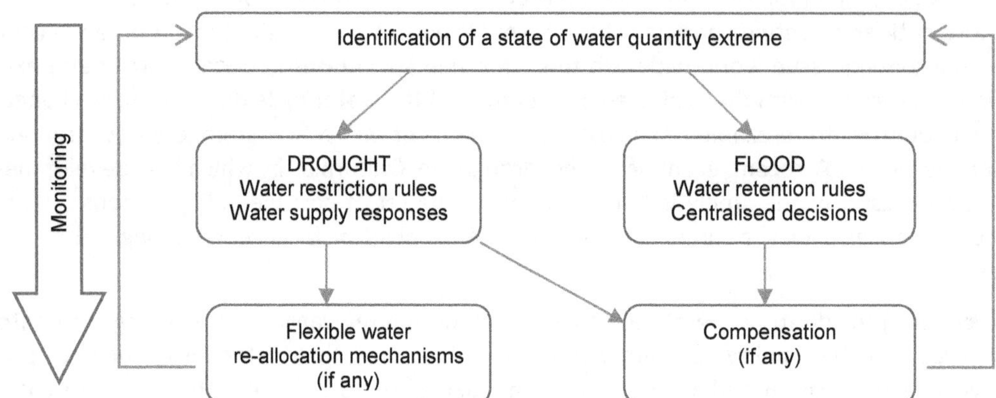

Figure 3.4 identifies three key steps of short-term water shortage policies during droughts. The first step consists in identifying the state of the water resource or flow. This requires monitoring tools and information systems, such as drought and flood indices, water level and flow monitoring, and a procedure to deal with this information and recognise the state of the water extreme (Step 1). In practice, this may involve stakeholder or the participation of experts, or more "automatic" rules based on monitoring systems (for example, the flow of a river).

If the state of water shortage is recognised as critical, then this can trigger a set of water restriction rules, eventually combined with short-term water supply responses, such as supporting a river's flow by releasing increasing volumes of water from a dam or allowing for groundwater pumping (Step 2). Short-term water restriction rules can concern all water users, but especially those that rely on freshwater withdrawals such as agriculture, urban and industrial users.

In certain cases, following or simultaneously to Step 2, flexible mechanisms for reallocating water across farmers or a broader set of users can be implemented (Table 3.1). In practice, this usually takes the form of water quota trading or water auctions by public agencies. In practice, Steps 2 and 3, i.e. water restriction rules and flexible water reallocation mechanisms, can take place simultaneously as two elements of a more global system for addressing temporary water shortages. In other cases, such reallocation mechanisms are not in place, so Step 3 does not exist. Finally, it can be also the case that water restriction rules, although non flexible, already intend to reallocate water as efficiently as possible (OECD, 2015a). Table 3.1 and Box 3.3 present examples of existing water restriction rules, depending on the initial water allocation system already in place for "normal" circumstances.

In circumstance of water shortages, the most common mechanism in most countries worldwide is to reduce water demand based on either a set of priority rules among water users and uses, or a temporary proportional reduction of water rights (OECD, 2015a). The two mechanisms can be put in place together. Typically, priority rules generally concern different water users and uses, such as agriculture, industry, and tourism, while a proportional reduction of rights are applied within the agricultural or irrigation sector. Another common mechanism to manage temporary water shortages is to forbid irrigation for a certain time period, i.e. a time quota, which can be considered an imperfect proxy for a reduction of water rights based on quantities.

Table 3.1. Water allocation systems and adjustment to droughts in agriculture

Water allocation system "Normal" circumstances	Water restriction rules "Extreme" circumstances (water shortage)	
	Restriction mechanism	Flexibility mechanism
Water rights (individual or group)	• Public buyback of water rights on spot markets • Public auctioning of water rights • Proportional reduction of individual or group water quota so that aggregate demand equalizes aggregate supply • Priority rule: priority users or group of users keep their initial quota; followers share the residual aggregate water supply.	– Water quotas trading – Water leasing
Administered water pricing	• State-dependent water scarcity pricing: prices vary with available water supply (rare in practice)[1] • Water pricing formula with increasing marginal price based on individual consumption (more frequent but not targeted to short-run adjustment to water shortages)[2]	
Regulation of irrigated areas	• Irrigation forbidden in certain areas for a given time period. • Combination of time restrictions with reductions of water quotas.	

1. Such administrative pricing formulas are seldom used in practice, for different reasons: difficulty of practical implementation, unintended redistributive effects; uncertainty in the outcome in terms of reduction of water demand due to unknown price elasticity of water demand, etc.

2. Such block-rate tariffs, when based on individual water consumptions, cannot be considered as restriction mechanisms *stricto sensu*, since they are not based on aggregate water supply. However, the fact they signal to individual water consumers an increasing marginal cost can provide incentives to adopt drought risk mitigation activities.

Interestingly, administrative pricing is seldom used to manage short-term water shortages, even if a pricing mechanism is in place. This can be explained by the fact that administrative water pricing in agriculture is rarely used, even in the long term, to reflect the opportunity cost of water, and thus allocate water across farmers, but rather as a financing tool for water infrastructures. Thus it is not surprising that it is not used much in the short term. Moreover, reducing water demand temporarily through a price increase requires some knowledge of the price elasticity of water demand, which can vary a great deal across time. In addition, water price increases may have negative equity implications for farmers, leading to regressive redistribution through water demand. Finally, political pressures from the agriculture sector are susceptible of undermining the development of such an instrument. Broader institutional aspects can also play a role. In the case of the United Kingdom such a variable administrative price scheme was proposed, and rejected for the reason that general tax policy in the country required that taxes be predictable and stable.

Box 3.3. Two examples of flexible water re-allocation mechanisms during water shortages

Volumetric management of water resources in France

An example of water management practice is volumetric management. Under this system, each farmer is allocated, for the growing season a water access right quota in volume terms, calculated on the basis of his crop portfolio in hectares. This global volume is then allocated across time periods, typically weeks, and it is forbidden to consume more water than the allocated volume during the set timeframe (e.g. week or decade). Volumes are estimated by a metering device at the farm level and can be controlled randomly by the Water Police at the end of each period. Water pricing for irrigation mainly covers operations and maintenance costs, and investment costs related to irrigation infrastructure. It is not used to reflect scarcity costs (OECD, 2010a). Irrigators must pay a contribution to the water basin agency, which can be differentiated between areas or irrigation techniques, but its average amount is usually considered insufficient as an incentive device to reduce water demand by irrigators (Lefebvre and Thoyer, 2012).

In times of water shortages, specific rules apply, that can be described as follows. First, the level of water availability is regularly measured by the regional offices of the French ministries of environment and of agriculture. Flows of rivers are measured at different points of the watershed, and compared to some predefined triggering alert thresholds. There are two levels of alert defined by trigger river flows: the summer ("*étiage*") flow trigger (EFT), and the crisis flow trigger (CFT). When water flow is higher than the EFT, there is no restriction for irrigation. If the water flow is between the EFT and the CFT, there may be restrictions, depending on the outcome of a standardised negotiation process involving stakeholders in the framework of a pre-defined drought committee. When the river flow falls under the CFT, then irrigation is forbidden in the watershed, in order to ensure a minimum environmental flow, aiming at ensuring the two priority uses that are tap water supply and aquatic life. These different restrictions are anyway temporary, and can be revised in light of the evolution of the state of water flow.

Water markets in Australia

Australia has been experiencing frequent and severe drought episodes in the last decade and has undertaken significant reforms of water management policies, especially in the agricultural sector. A specific feature is the use of water markets. Farmers are allocated water entitlements each year which can be revised according to the state of water reserves and rainfall. There are two distinct water markets: a long-term water market, and a short-term water market. Each is targeted to a specific timeframe. Short-term water markets allow farmers temporary water exchanges, allowing for a more efficient allocation of water within the growing season, and so a reduction of the overall cost for farmers as whole.

The interesting specificity of the Australian water markets is that individual rights are defined as a share of the total volume of water available for a given period of time. Government buys back water rights to ensure that enough water is available for a proper functioning of ecosystems (minimum environmental flows). Within the irrigated sector, water trading allows the system to continuously and almost automatically adapt to changing weather and hydrological circumstances during the growing season. A second feature of the system is that these shares can be exchanged between farmers. These two features taken together allow to significantly decrease the overall cost of droughts for the farm sector as a whole, as shown by Mallawaarachchi and Foster (2009) in a study covering the 2007-2008 water shortage in Australia.

Source: Extract from OECD (2014), http://dx.doi.org/10.1787/9789264209138-en.

Despite these limitations, some irrigated perimeters or irrigation companies continue to use some form of administrative water pricing mechanisms to reduce the risk of water shortage. Typical examples include increasing block-rate tariffs, which provide an imperfect signal to water users of the increasing opportunity cost of water scarcity, or more refined nonlinear pricing scheme. Block-rate tariffs are a water formula pricing based on different layers of water consumption amounts, each layer being affected a unit price of water. Under increasing block-rate tariffs, the unit price of water increases as the volume of water increases from a layer (a "block") to the following one. Sidibé et al. (2012) have shown, using a simulation model, that such nonlinear pricing systems allows to mitigate with a certain efficacy the negative impacts of droughts for risk averse farmers in the case of two irrigation companies in France: the *Compagnie d'Aménagement des Côteaux de* Gascogne and the *Compagnie d'Aménagement des Eaux des Deux-Sèvres*. Still, there are limitations when tariffs are based ultimately on individual water withdrawals since they do not reflect the overall degree of water shortage. When feasible, a mechanism based on aggregate water availability and demand such as the Australian water markets is likely to yield more efficiency gains.

The dominance of temporary priority rules and proportional right reductions in times of water shortages raises the issue of their efficacy and efficiency, and equity. If the objective is to equalize marginal benefits from water uses across users, such rules are clearly imperfect. Priority rules may not properly reflect the relative (marginal) value of water across users and uses, while a proportional reduction of water rights may be more damaging for certain water users than for others. For example, farmers with perennial crops compared to farmers with annual crops. A solution to improve the efficiency of water allocation in the short-term is to allow for temporary or seasonal water-trading. Short-term water trading has been experienced in Australia for years, and assessments of the benefits of the system tend to show that it has led to substantial gains in economic efficiency, especially during droughts. There is a vast amount of literature on Australia in this area that strongly suggests that water trading, especially on short-term markets, not only allows to substantially mitigate the impacts of droughts on agriculture, but is now considered by farmers as a useful and basic instrument in their risk management toolbox. The policy challenges for Australia in the area of water regulation in agriculture is more in the areas of how to fine-tune the existing system and robustness issues related to climate change.

Fine-tuning vs simplicity? While the Australian example illustrates the potential efficiency gains arising from water trading, especially for managing water shortages in drought circumstances, the choice of an instrument must always consider the full range of costs and benefits. Running a short-term water market requires investment, at least at the initial stage, in water monitoring systems, hydrological knowledge, skills, institutional changes, confidence among others. In countries where severe to extreme droughts are relatively frequent, the efficiency gains of running such markets are likely to outweigh the costs. In countries with less frequent drought episodes or with less structural sensitivity to water shortages, simple priority rules or proportional reductions of water rights may in fact represent a reasonable second best solution. However, climate change as well as market and policy drivers may relatively quickly affect the cost-benefit ratio of different approaches, so it seems important that countries when reforming their water policies to reduce vulnerability to drought risks, consider the value option of systems able to evolve towards more flexibility in the future, and avoid lock-in effects. The Australian example is also useful to consider regarding this dimension, since the reform process involved costly buy-backs of water rights.

Beyond the different formal mechanisms in place to quickly adjust water demand in supply during water shortages (priority use, proportional reduction of rights, markets), one should not understate the importance of information production and dissemination — meteorological, hydrological, and agronomic — and governance processes. Firstly, information allows farmers to revise their production plans during the drought onset. Although estimates of the benefits of such use of information are sparse, one can reasonably think they are important in farm systems that are not too specialised or dependent on water resources (Reynaud, 2009). Secondly, decision-making regarding water restriction rules, whatever the country considered, always involve a mixture of pre-defined rules (alert thresholds, minimum environmental flows) based on scientific evidence and negotiation process among water users. There is here the classic question of *rules vs discretion* due to the potential lack of credible commitment of water authorities and the strategic use of *fait accompli* in the negotiation process (Kydland and Prescott, 1977).

A final issue for the management of water shortages is the allocation of the cost of water restrictions. In countries like France, temporary restrictions of irrigation or reductions of water rights related to drought episodes do not lead to government compensation from the authorities in charge of water management. Water is considered as the common property of the nation, thus water rights provided to farmers can be reduced by the initiative of the government (in practice its local representative is the *Préfet*). Farmers do not own water rights *per se*; they have a right to use water provided by the government. A polar case is Australia where the government is buying backwater entitlements[1] from farmers to ensure equilibrium between water demand and supply. This, *de facto,* provides compensation to farmers for the reduction in water rights. Although this has contributed to increase the acceptability of the reform by irrigators, it raises the question of the cost of the programme in addition to equity issues.

3.2. Water risk mitigation policies: Floods

Agriculture, green infrastructures and flood risk mitigation

Although floods are a significant risk for agriculture in several regions of the world, there is comparatively less literature, at least in the area of water economics, on the impacts of flood risk on agriculture and flood risk mitigation strategies. From a technical perspective, managing flood risks is not a new issue for farmers. Flood risk is not always a cost for farmers: historically, some of the most fertile farming systems at the beginning of agriculture were associated with river systems, such as the Nile system, which allowed for regular flooding of agricultural land and renewing of soil fertility.

Today, at the farm level, changing land uses or using agricultural practices that favour water infiltration, machinery protection and technical capital can reduce exposure to floods. This menu of practices can be affected by the flood risk when farmers are risk averse and should be accounted for in land use choices: the most exposed plots would be used for less valued crops (PREEMPT, 2012).

Flood risk policy measures can be divided into two different groups: structural measures and non-structural measures. Structural measures (e.g. dams, dikes) interfere in the phenomena of flood formation and routing; and non-structural measures foresee, prevent and mitigate the impacts of flooding (European Commission, 2010). From a historical perspective, with a certain analogy with drought risk, structural measures were the main instrument used in the past to reduce flood risk. However, the cost-benefit ratio of structural measures could be challenged by the rising uncertainty about projected rainfall patterns due to climate change (Foudi, 2013). Currently, flood management plans are more focused on preparedness, monitoring and prevention actions.

In an analogy with water shortage risks, flood risks can broadly be considered as a "common bad" from an economic perspective, thus an efficient flood risk mitigation strategy must be considered in an integrated manner, beyond economic sectors and administrative areas. This is notably the case of Integrated Flood Management (IFM), an approach promoted by both the World Meteorological Organization and Global Water Partnership since the early 2000s, building on the pre-existing concepts of Integrated Water Resource Management (IWRM). The principle is to "integrate land and water resource development in a river basin, within the context of Integrated Water Resource Management, with a view to maximising the efficient use of floodplains and to minimise loss of life and property" (World Meteorological Organization, 2009). In the European Union, the trend is towards prioritizing from "grey" to "green" infrastructures in close relationship with the concept of IFM, with a particular emphasis on the development of Natural Water Retention Measures (NWRMs) (European Commission, 2013; Linnerooth-Bayer et al., 2013).[2] At national levels, typical examples include the Room for River initiative in Netherlands.

According to Kenyon et al. (2008), "there are currently few institutional links between flood risk management and agriculture, there is great potential for agriculture to become part of the solution to flood risk management rather than being part of the problem." Agricultural land management practices change the hydrological properties of a surface water catchment. Thus, these practices have the capacity to increase flood risks, as well as provide potential flood mitigation and protection to areas downstream (Kenyon et al., 2008; Morris et al., 2010; Schilling et al., 2013). There are several measures for delaying runoff to mitigate local flooding, including using grass buffers, temporary ponds, and ditching (Environment Agency, 2007). The restoration of floodplains and wetlands can store water in periods of high or excessive precipitation for use in periods of scarcity. The European Commission (2011)[3] proposed different natural flood management methods that are summarised in Table 3.2. Techniques that are related to the agricultural sector are presented in Table 3.3.

A major challenge to such an approach, however, is that the relationship between land management and flood risk is complex and highly site-specific (Pattison and Lane, 2011). In particular, even if the knowledge of the causes of local flooding are reasonable, it is difficult to upscale at the water basin level to provide a global picture of how changes in land use or farm practices can affect flood risk at this scale (O'Connell et al., 2007). This lack of scientific evidence is a practical challenge for policy making.

Table 3.2. Techniques of natural approaches to reduce flooding

Technique	Application	Potential location	Key goals	Notes
Hedgerow planting and management	Flood plain, Catchment-wide	Planted across-slope along existing field boundaries	• Enhance infiltration and storage • Impede overland flow of water and sediments	Perhaps suited to more intensive agricultural landscapes
Channel reprofiling	Riparian	Creating a two-stage channel	• Maintain adequate depths during low flows • Enhance winter storage • Encourage more natural morphology	Requires consultation with statutory bodies
Blocking of inappropriate artificial drains using dams	Riparian	Any artificial drain throughout catchment, provided it would not increase flood risk	• Slow flows • Enhance water storage • Intercept excess sediments	Blockages may be constructed by large woody debris, earth, rocks, bales of hay or heather. Plastic piling may be required if incisions reaches mineral substrates
Wetland restoration	Riparian, floodplain, catchment-wide	Flat upland areas, foothills and floodplains prone to waterlogging	• Enhance flood storage capacity throughout the catchment	Can be online (i.e. physically linked to watercourse) or offline (e.g. on flat hilltops)
Gully woodland planting	Riparian, catchment-wide	Upland gullies	• Impede rapid runoff entering steep channels • Contribute LWD to channel	May require livestock fencing
Native mixed woodland on hillslopes	catchment-wide	Deforested and drained hillslopes	• Intercept rainfall • Enhance soil storage capacity • Reduce erosion if actual soil coverage is effective	Planting on north-facing slopes, gullies can enhance snow-pack retention, desynchronizing winter flood peaks
Floodplain "leaky barriers"	Floodplain	Key floodplain zones (not close to buildings or important infrastructures)	• Intercept overland flows • Enhance floodplain storage potential for both water and sediments	Living walls of woven willow spilling can be constructed to disrupt flow paths over floodplains
Planting riparian buffer zones, or water margins	Riparian	All water courses, particularly heavily modified watercourses and those within artificially drained areas	• Impede overland flow • Enhance soil storage capacity • Intercept mobilised debris and sediments	May require fencing and provision of alternative water sources for livestock

Source: European Commission (2011), http://ec.europa.eu/environment/water/flood_risk/better_options.htm.

Table 3.3. Runoff mitigation measures in agriculture

Response theme	Measure	Examples
Water retention through management of infiltration into the catchment	• Arable land use practices • Livestock land practices • Tillage Practices • Field drainage (to increase storage) • Buffer strips and buffering zones • Machinery management	• Spring cropping (versus winter cropping), use of cover crops. • Extensification, set-aside and arable reversion to grassland. Lower stocking rates, , restriction of the grazing season • Conservation tillage, cross-slope ploughing • Deep cultivations and drainage, to reduce impermeability Contour grass strips, hedges, shelter belts, bunds, riparian buffer strips • Low ground pressures, avoiding wet conditions
Water retention through catchment-storage schemes	• Upland water retention • Water storage areas	• Farm ponds, ditches, wetlands • Washlands, polders, reservoirs
Managing conveyance	• Management of hillslope connectivity • Channel maintenance • Channel realignment	• Blockage of farm ditches and moorland grips • Reduced maintenance of farm ditches

Source: Morris et al. (2010), http://dx.doi.org/10.1787/9789264083578-9-en.

Flood crisis management and water storage on agricultural land areas

Flood crisis management is a complex topic involving institutional, planning, economic, and sociological dimensions that are beyond the scope of the present report. As potential victims of flooding, farmers are part of general flood risk policies, including its flood crisis management dimension. There is, however, one aspect for which agricultural land areas is likely to play a significant role in the course of flood crisis management: temporary water storage, or floodplains.

Agricultural land areas may have the function of flood water storage, notably because they represent significant shares of total land areas in most countries of the world. The basic rationale for floodplains is that the value of flood damages can be lower on agricultural lands than on other lands, such as urban areas where wealth is more concentrated per unit area and damages from floods can be long-lasting. This would allow to reduce the overall cost of flood damages for society, and thus improve social welfare. These differential costs are not just of a material nature, they also concern irreplaceable commodities, such as historical artworks, architectural ensembles, and lives saved. The cost differential should also include the recovery phase which can be substantial. From an economic perspective, temporary water storage on agricultural land areas can be considered as a way to improve the efficiency of water allocation considered as a public bad, in analogy with the optimal allocation of water under shortage circumstances.

This potential benefit of flood water storage on agricultural land areas should not be considered as a substitute to land planning and the regulation of urbanisation, but rather as an additional management instrument in a comprehensive toolbox in an integrated management of flood risk. Increasing risks of floods in urban areas are strongly linked to the increasing pressure of urbanisation, sometimes in areas considered at serious risk of flooding.

In relation to the priority placed on the development of structural measures in the second half of the 20th century, the use of floodplains have been to a certain extent put aside until recently, except in certain countries. Expansion of urban areas in OECD countries is potentially creating new zones of risk exposure, sometimes in areas that were considered hundred years ago as natural floodplains (for example, in the area of the city of Lyon in France, *cf.* Combe, 2004). Still, there are several examples of flood water storage on farmland in OECD countries (see Erdlenbruch et al., 2009 for France). For example, in the United Kingdom (Box 3.4), Beckingham Marshes is a 900 ha area that has had a storage function of flood waters for more than half a century which regulates the flow of the River Trent, near the town of Gainsborough. This reduces the impact of flooding *vis-à-vis* urban areas and contributes to minimising the overall economic cost. Technical infrastructure pumping and drainage can control more precisely the function of flood control. Farmers are not the majority owners of this area, which belongs mainly to the Environment Agency (Morris et al., 2010). In Japan, paddy field and waterways have a function to store rainwater temporarily and can contribute to reducing the damage of extreme events to farm and residential land.

Box 3.4. Examples of integrated flood risk management initiatives: Netherlands and the United Kingdom

Netherlands

In the Netherlands, the *Deltaplan Agricultural Water Management*, an initiative of LTO Netherlands (Dutch Federation of Agriculture and Horticulture) in close collaboration with the Ministries of Infrastructure and Environment and Economic Affairs, provinces and the drinking-water sector, aims to respond to the water challenges of the coming decades that entail a significant role for agriculture. As a considerable part of the agricultural and water challenges go beyond the competence and financial capacity of regions, structure, integrality and connectivity are needed. The initiative covers water quantity and quality problems, as well as land planning with a view to making "space for water", including: i) an assessment system that encourages the economical use of space and minimises the loss of agricultural land for other functions; ii) an agricultural compensation arrangement in the case of the transformation of agricultural land to other functions; and iii) optimisation of management schemes with which farms can supply public/societal services. In terms of water quantity, the initiative comprises notably a higher degree of self-sufficiency on farm and regional levels by new innovations for water saving at the farm level and conservation at the regional level.

United Kingdom

The role of agricultural land in flood risk management has been important in the United Kingdom. Several initiatives and pilot projects have been conducted that include agricultural lands as key players of flood management programmes, in the context of the programme Making Space for Water. One example is the recent pilot project Payment for Ecosystem Services (PES) on Flood Regulation in Hull. The objectives of this pilot programme were to characterise the current state of ecosystem services delivered to urban areas; to identify potential improvements of ES delivery; and to design potential payments for ecosystem services that could be applied. Two initiatives followed from this programme: PES schemes, including a country Park scale PES which would allow mitigating flood risk in the north-west of the Hull by developing "swales, bunds, ponds, replacement of permeable road and car park surfaces and conversion of amenity grassland to semi natural grasslands and more varied woodlands". The second initiative is the Beckingham Marshes Washland Creation which aims to create 94 hectares of floodplain grassland in order to improve flood risk mitigation for the towns of Gainsborough and Beckingham on the River Trent. While agricultural land areas used to contribute to flood risk management in this area since the 1960's, this project also addresses restoration of natural habitats.

Source: UK Environmental Agency (2010), http://webarchive.nationalarchives.gov.uk/20140328084622/http:/cdn.environment-agency.gov.uk/geho0310bsfi-e-e.pdf; URSUS Consulting (2013), www.google.fr/url?sa=t&rct=j&q=&esrc=s&source=web&cd=1&ved=0ahUKEwjAqobl4tvJAhVMtBoKHXnoC8UQFggcMAA&url=http%3A%2F%2Frandd.defra.gov.uk%2FDocument.aspx%3FDocument%3D11136_FinalreportHullPESPilot130513.pdf&usg=AFQjCNFMCsrXZavhwJ43r31pRcAMueHTiQ&bvm=bv.109910813,d.ZWU

The 2007 European Directive on the Assessment and Management of Floods explicitly incorporates this dimension of natural flood risk management. In France, flood prevention action programmes are designed to mitigate the costs of floods by means of a set of planning and incentive tools. The possibility to transfer flood risks from high potential damages areas (urban and industrial) to lower damages areas such as agricultural land is part of an overall effort to mitigate flood risks (Erdlenbruch et al., 2009). The implementation of these plans has revealed several issues, including the need for an objective and transparent definition of flood risk, and the use of a compensation mechanism and/or insurance to compensate for economic losses of farmers whose land was flooded.

Which policy tools to foster the role of agriculture in flood risk mitigation?

If the role of agricultural land areas in integrated flood risk management is increasingly recognised in different OECD countries as a potentially efficient and lower cost option than structural measures, which policy tools to foster in such a role has not yet been defined. In theory, the role played by floodplains or the adoption of *ex ante* land use and agronomic practices can be considered as a form of ecosystem service, which makes the specific flood risk mitigation problem a subset of a more general discussion on agri-environmental policies and ecosystem services. However, there is more complexity in this area since countries have also specific flood risk policies already in place, with their own logic and institutional history. There is thus a question of policy coherence between agri-environmental policies, which can include flood risk mitigation objectives, and more general flood risk policies.

Building on the premise that flood risk mitigation by agriculture can be considered as an ecosystem service, there are challenges in the design of agri-environmental policy tools that are common with other agri-environmental issues. These include: asymmetric information between the regulator and farmers; uncertainty about the outcome of the measure; and the need for scientific data to find a good proxy between practices and outcomes. These challenges have been analysed in more depth in the OECD *Guidelines for Cost-effective Agri-environmental Measures*, with several recommendations that are sufficiently general to apply to the domain of flood risk mitigation (OECD, 2010b).

The biggest challenge for designing incentives to foster farm practices able to reduce flood risks is to overcome the lack of scientific knowledge between these practices and outcomes. Uncertainty in outcome is certainly undesirable for objectives such as bird conservation, but becomes socially unacceptable when it is about flood risks that can cause huge damages in cities, as well as injuries and loss of life. This may appear to give the advantage to structural measures which, despite their limitations and defaults, could offer more safety. However, this question is fundamentally site-specific. One can hardly imagine a successful and socially acceptable policy strategy fostering green infrastructure that would not be able to deal in a reasonable way with outcome uncertainty.

Outcome uncertainty is perhaps less an issue for floodplains, as they are mostly dependent on local context and engineering works. There is no generic database on the use of floodplains in OECD countries, but the practice is not new and seems to be a growing element of general flood risk policies, for example in the United States or the European Union. According to the *OECD Member Country Questionnaire Responses on Agricultural Water Resource Management* (OECD, 2009), many OECD countries have economic incentive measures for water retention and storage (Annex 3A.1, Table 3.A1.2). Economic incentives for wetlands are implemented by seven OECD countries, and agri-environmental schemes can include a dimension of flood management. In sum, economic incentive measures appear to be well represented, but in the absence of a more detailed description of their content, it is not possible to compare them or to draw more precise conclusions.

The diversity across OECD countries could also come from the fact that drought and flood risks have different characteristics across OECD countries in terms of frequency, magnitude and economic impacts. This may affect the costs and benefits, including implementation and transaction costs, of a given policy tool. To address the relatively infrequent and limited magnitude of droughts, simple regulatory tools can be sufficient. If droughts are more frequent and have huge impacts, more sophisticated tools would be desirable.

In terms of cost-effectiveness and policy design, several recent studies have underlined the potential for floodplain conservation or natural water retention measures as socially efficient (from an economic perspective) flood-risk mitigation response. For example, using a combination of data and modelling tools (e.g. a flood model, geographical information data) in the Meramec River in Saint-Louis County, Missouri, Kousky and Walls (2014) show that the benefits of floodplain conservation, including avoided flood losses and home prices capitalisation, outweigh their opportunity cost by a ratio of 1.8. However, they also show that avoided flood losses considered in isolation are not sufficient to ensure a positive benefit-cost outcome. Another recent study conducted for the European Union estimates the cost-effectiveness of different water retention measures, and show that crop practices and grasslands are susceptible to be the best option to reduce peak flows in different European countries such as France, Spain, southern Italy and Greece (Burek et al., 2012).

Cost-effectiveness studies suggest an economic case for floodplains, but also that they could be considered not in isolation but in a broader bundle of ecosystem services. In many cases, floodplains do not just affect flood risk, but have additional outcomes in terms of biodiversity and landscape. Not only would these elements be usefully included in cost-benefit analysis, but also in the design of agri-environmental policy tools targeting a bundle of environmental goods. This raises the issue of additionality of ecosystem services and potential stacking of ecosystem services payments, on which the OECD has recently devoted work (Lankoski et al., 2015) that may be helpful to improve the design of agri-environmental policies, including flood risk mitigation objectives.

3.3. Compensation and insurance policies against droughts and floods

Compensation policies constitute the last category of policy tools available to cope with droughts and floods. The objective is to compensate, mostly in financial terms, people impacted by such events. Beyond this general objective, compensation policies can take very different forms in practice: from pure private insurance markets with insurance premiums reflecting risks and freedom to insure or not, to state insurance through public funds with mandatory participation. Almost all OECD countries have compensation policies against droughts and floods (OECD, 2009).

As underlined in the previous section, there is a general issue of insurability. Drought and flood risks have specific features that can make them difficult to insure at a "reasonable" cost (i.e. willingness to pay is significantly higher than the insurance premium). Such problems, associated with a real demand for compensation tools, have led many OECD countries to develop policy schemes to overcome the absence of coverage against these risks. The relative efficiency of the different schemes observed in OECD countries is subject to considerable policy debate and has been extensively addressed in OECD work on risk management (OECD, 2011). A robust outcome from this work is that one must consider different risk layers for which specific instruments are required. Extreme cases of droughts and floods most often belong to the category of catastrophic risks, and thus require some form of government intervention.

The boundaries between "catastrophic", "insurable" and "normal" risks cannot be reduced to the space delineated by probabilities, consequences, and spatial correlations. An efficient definition of the boundaries between catastrophic and non-catastrophic risk is a challenging exercise in the context of climate change that tends to increase the variance in water availability (OECD, 2014). The boundaries become a moving target in the context of non-stationary climate. In practice, the boundaries of catastrophic risk are defined to a certain extent by policy, and they have implications for structural adjustment and climate change adaptation. Policy action, in particular payments and compensations after floods and droughts, will influence the risk management strategies followed by farmers. In particular, farmers are likely to perceive that the part of the risk covered by the government is no longer their own responsibility. Policies tend to crowd out risk management and adaptation strategies such as diversification or structural adjustment of the activities in the farm (OECD, 2011).

From ad hoc to ex ante risk management approaches

There are differences in OECD country approaches for the allocation of risks considered as catastrophic and uninsurable. Countries such as Australia, Italy and New Zealand consider that farmers, to a certain extent, must include the whole set of risks they face in their business operations in an integrated risk management strategy, including extreme weather events, and should not in principle rely on government for support (Box 3.5). In Australia, the focus of national drought programmes has shifted in recent years to preparedness, risk management, self-reliance and resilience. For the past two decades, provision of drought support had relied on the declaration of "exceptional circumstances" (EC), involving support measures such as grants, interest rate subsidies and income support payments—as well as several training programmes and counselling services. Following a major national drought policy review in 2008 and 2009, the Australian government agreed that the EC declaration approach was no longer appropriate in the face of a variable climate. In 2013, the Intergovernmental Agreement on National Drought Program Reform (IGA) was approved between the Australian, state and territory governments, underpinned by the 1992 National Drought Policy. The agreement aims to assist farmers to prepare for and manage climate variability, including drought, and adopt self-reliant approaches to manage their business risk rather than waiting until they are in a crisis to offer assistance. The IGA focuses on five key outputs: i) an income support payment for farmers and their partners in hardship (not drought-specific), ii) continued access to taxation reduction measures and the Farm Management Deposits Scheme (which enables farmers to put aside money in good years and draw from it in years of lower income), iii) a national approach to farm business training, iv) a coordinated, collaborative approach to the provision of social support services, and v) tools and technologies to assist farmer decision making.

In Australia, drought is considered as an important issue for sustainable productivity growth in agriculture. The Australian government is currently reviewing a new drought policy approach in terms of the adequacy of current guidelines relating to drought preparedness and in-drought support as part of the development of the Agricultural Competitiveness White Paper process.

Until recently, France considered that such catastrophic events, by nature unpredictable, should be considered as a matter of national solidarity, rather than an element of risk management. However, France has undertaken significant reforms of its agricultural insurance system since the mid-2000s (Mortemousque, 2007). Before 2005, only hail risk was covered by private insurance markets with no or only limited government support. The *Fond National de Garantie des Calamités Agricoles* (now the *Fonds National de Garantie des Risques en Agriculture*) used to cover risks considered as non-insurable, including droughts and floods. The fund is financed by farmers and, if necessary, by the State. Compensation is triggered by an official declaration of an agricultural disaster (Babusiaux, 2000). Since 2005, insurance markets have developed multi-risks insurance contracts, including drought risk for cereals and private multi-risks insurance contracts are subsidised by the State and the European Union, up to 65% of the insurance premium. Experiments are also underway to develop a potential weather index insurance tool for pastures. Risks for which insurance has reached a sufficient level of development are excluded from the Fund.

The boundaries are not easy to define because the optimal design depends on the relative transaction costs of different institutional arrangements in water systems. In a certain way, boundaries are not just a question of probabilities, they are also endogenously determined by the relative transaction costs of alternative policy instruments. These boundaries can shift with different investments in infrastructure, and different organisational and institutional details of each instrument.

Box 3.5. Compensation against weather risks in agriculture: New Zealand

New Zealand pasture-based livestock farming systems are highly susceptible to varying climate events. The two largest flood or storm events experienced by New Zealand were Cyclone Bola (1998) and the Southern North Island Floods (2004). In 2013, New Zealand experienced the worst drought in 40 years affecting the whole of the North Island and parts of the South Island's West Coast. The estimated cost to the economy was NZD 2 billon, reducing GDP in 2013 by an estimated 0.7%. In response to droughts, land owners are changing farm management practices by increasing the use of supplementary feed and use of more drought resistant pastures and crops. Climate change is increasingly considered a major issue, with more extensive areas of agricultural land projected to be prone to droughts and flooding, even in areas that have not been subject to these events up to now.

In New Zealand, the development and implementation of policies encourage farmers and landowners to take responsibility for managing and mitigating these risks to their businesses. Policies are set in such a way that direct government support is only provided when the scale and impact is beyond what farmers could be reasonably expected to either cope with or mitigate against.

The New Zealand government does not intervene financially in agriculture insurance for climate based risk. However, it provides legislation to support mutual insurance cover: Farmers' Mutual Group Act 2007. Moreover, the Inland Revenue Department runs an income equalisation scheme to allow farmers (taxpayers) to even out fluctuations in income by spreading their gross income from year to year. The adverse event income equalisation scheme allows farmers to carry income from forced livestock sales over to the next income year.

The *New Zealand Primary Sector Adverse Events Policy* sets out individual, community and government responsibilities across a scale step classification of events based on impact and area effected (localised, medium, large scale). In a large scale event a Special Recovery Measure (SRM) may be enacted. The SRM is targeted at restoration of uninsurable assets only: uninsurable infrastructure, uninsurable silt and debris removal, uninsurable pasture, crops and forestry plantations (excluding slips). The reimbursement rate is capped at 50%. Insurable assets are not covered, the primary responsibility for recovery lies with individuals.

The New Zealand government can also provide financial assistance to the Rural Support Trusts following a medium or large scale adverse weather or natural disaster event. Rural Support Trusts are a nationwide network that helps rural people and their families during and after extreme weather or environmental events which affect their livelihoods. This includes pastoral farming, forestry, horticulture and other land based activities. Many of these trusts offer their services in times of general hardship.

Source: New Zealand Ministry for Primary Industry, www.mpi.govt.nz (2014).

Recognising the formal superiority of *ex ante* approaches based on the principles of risk management, notably on the basis that it provides more visibility for farmers compared to *ad hoc* compensation schemes, many OECD countries have developed public-private partnerships to develop agricultural insurance systems, which in most cases involve different forms of financial support: premium subsidies (France, Spain, United States), administrative cost-sharing schemes, or public reinsurance of last resort (Spain, United States). Although there are significant differences across OECD countries as regards their agricultural insurance policies, the basic underlying idea is that, if it is necessary to subsidise to allow such compensation schemes to exist, it is preferable to subsidise *ex ante* than *ex post*. This logic has two major limitations.

First, when crop insurance premiums are subsidised, this means they only reflect a share of the risk, and thus may influence farmers' production choices and risk exposures. Compared to a situation with no subsidy and missing insurance markets, crop insurance subsidies may notably increase the demand for risk-increasing input such as fertilizers, while decreasing the demand for risk-decreasing inputs such as pesticides. Crop insurance subsidies may also affect crop rotations, with induced impacts on environmental performances greenhouse gas. Crop insurance subsidies are also susceptible of inciting farmers to expand their crops to marginal lands, leading to potential increases in vulnerability to drought risk (Claassen et al., 2011). In several OECD countries, crop insurance subsidies are a strong incentive element to buy crop insurance, beyond the reduction of the cost of risk itself. If some form of government support is necessary in this area, its level should be limited to a level where does not artificially boost insurance demand beyond the primary motive of the intervention, which is to foster a more efficient allocation of risks in the economy.

Secondly, it may be the case in practice that *ex ante* crop insurance programmes may be unable to crowd-out *ex post, ad hoc* compensation programmes. This has been observed, for example, in the United States where crop insurance subsidies coexisted with the option of disaster bills (although it is notable that there were no disaster bills for the 2012 Midwest drought or the 2013 California drought). or more recently in Australia in the case of interest rate subsidies following extreme drought conditions. This issue is a typical time consistency problem due to a lack of credible commitment to bail out farmers once the state of nature is realised and if drought impacts appear to be substantial enough to threaten the continuity of business operations by farmers. *Ex ante* approaches, based on crop insurance subsidises, were believed to be a rational response to such commitment problem; however casual evidences suggest the problem looks more complex. Overall one can consider that the challenge of crop insurance is not just a risk allocation problem but is at the interface of the areas of risk management, theory of industrial regulation (Tirole, 1988), and political economy.

Other approaches to address risk include financial tools dedicated to precautionary savings either through individualized or mutualised systems. Precautionary savings are defined as the extra savings made by an economic agent due to the presence of a given risk in the future. When insurance provided by markets is too costly, precautionary savings can be a cheaper substitute risk management tool. They also allow to manage all risks whatever their origins: weather, markets, accidents, etc. Of course, there are challenges with this approach: uncertainty on risk assessment remains and building precautionary savings can be difficult or impossible for those who are in financial stress. Several OECD countries have encouraged precautionary savings as risk management tools, notably Canada and France. Lessons from these countries suggest that the adoption of such tools by farmers is limited without additional incentives such as subsidies. In some cases, precautionary saving programmes have been used for other purposes, such as retirement. There is neither systematic data on the use of precautionary saving tools by farmers in OECD countries, nor on government support in this area. However, there is limited rationale on the ground of public economics to justify government support in this area, except on the ground of farmers' education and training.

Mutualising precautionary savings across farmers or regions can be done through mutual funds that provide compensations following a weather event up to the limits of the funds available. Although compensation may be insufficient compared to farmers' realised losses, it may constitute a cost-effective precautionary saving tool if administrative costs are lower than those of more complex insurance products. A major issue for mutual funds, however, is the risk of cross-subsidies if individual farmers' contributions are not calculated on an actuarially fair basis. Hence, the system may be unstable and require mandatory participation, as in the case of the French *Fonds National de Garantie des Risques Agricoles* (FNGRA).

There is a lack of assessment of the comparative impact of existing mutualised precautionary savings systems and private insurance market supported by government subsidies on farmers' actual propensity to engage in riskier investments and production. Undertaking such a comparative assessment would require having information on the incentives each system generates on farmers' decisions.

Opportunities to complete missing insurance markets against droughts and floods

As regards compensation policies, flooding is typically included as an insured peril in multi-peril crop insurance together with all other perils which may cause yield reduction. Providing agricultural flood insurance faces a number of challenges, such as the delineation of losses caused by flooding; quantification and underwriting of flood risks and financial challenges related to risk transfers and reinsurance, and perhaps more serious problems of moral hazard and adverse selection (World Bank, 2010). Long-term losses due to business interruption and degradation of soil quality, which can reduce soil productivity in the longer term, are difficult to include in insurance schemes. However, completing insurance contracts could provide good opportunities to improve farmers' coverage against floods.

As regards drought insurance, there is now a substantial set of pilot experiments on the use of weather-index insurances, i.e. insurance schemes with indemnification triggers based on the realisation of a certain value of a weather variable such as rainfall of temperature, instead of the real loss incurred by the farmer. Potential benefits and inconveniences of these insurance schemes have been extensively discussed. The main conclusion is that they are supposed to be less costly to implement (less monitoring and control costs, less moral hazard), with the drawback that they would insure an imperfect proxy of farmers' real losses (which leaves what is called the "basis risk" to farmers). Most pilot experiments take place in developing countries, in part due to their potentially more favourable cost-efficiency ratio. There is certainly room for OECD countries in this area to take stock of these pilot case studies to see if such weather-index insurance schemes could be part of their agricultural insurance programmes. Such tools may also facilitate the development of private reinsurance against spatially correlated risks, which is absolutely necessary to ensure insurability at an affordable cost for farmers.

Innovations based on satellite imagery could also provide an opportunity to improve the coverage of pastures against droughts. Pastures represent on average 67% of total agricultural land area in OECD countries, hence expanding forage insurance may provide useful value added to the sector. In France, a private crop insurance firm, Pacifica, has proposed since 2011 a forage insurance offer, in cooperation with Astrium, a 100% subsidiary of Airbus Group, European leader in satellites, and the tele-detection laboratory of the engineer school of Purpan located in Toulouse.

3.4. Comparing policy approaches to droughts and floods in agriculture in Australia, Canada, France, Spain, and United Kingdom

The objective of this section is to present and analyse from a policy perspective the main similarities and differences between policy approaches to droughts and floods in agriculture in five OECD countries: Australia, Canada, France, Spain, and United Kingdom, on the basis of a review undertaken by OECD and to draw possible policy lessons. The main conclusions of this benchmarking exercise, essentially of a qualitative nature, are summarised in the following. The synoptic tables presented in Annex 3.A2 allow a comparison of the main elements and characteristics of drought and flood policies in agriculture in these five countries regarding their components: long-run water risk mitigation policies and prevention, short-run water management policies; and compensation policies. These tables also provide information about country approaches in terms of risk-thresholds and layering for both droughts and floods.

Drought policies: The benefits of market-based approaches

In spite of very different climatic conditions, there are similarities between Australia, Canada, France, Spain and United Kingdom in terms of drought risk mitigation policies. First, all five countries exercise control on the use of water by farmers, with legal mechanisms to curtail water use rights, and use centralised allocation mechanisms in times of shortages.

For all five countries, improving the technical efficiency of irrigation water use at all levels of the water systems has been a major ongoing policy effort. At the same time, in all countries, different pricing mechanisms are established, but prices rarely ensure complete cost recovery. This is not specific to the countries considered, but rather a general observation in the OECD area. Some irrigation water management companies in France have developed water pricing methods that are based on a previous water subscription before the agricultural season. These pricing systems allow to forecast water demand, which leads to a better management of droughts. Spain's water charges vary significantly across basins and even districts. The energy component of water use has become a strong deterrent for more than half of the irrigated area (using groundwater or modernised irrigation schemes).

All five countries have advanced monitoring systems and information tools. Some of them are specially targeted to the agricultural sector, highlighting the value of this information for farmers, and the possibility to adapt their production plans in the course of the growing season to mitigate the impacts of droughts on their farming systems.

Australia, Canada, France, Spain and the United Kingdom distinguish normal conditions from "crisis" circumstances, but they use scales of severity and risks rather than a binary approach and implement different measures to operate in one or another situation. None of the five countries have drawn a unique boundary to establish when crisis or normal conditions prevail. Rather, they have defined sets of scores or thresholds with different stages of severity. Based on these, different policies are implemented, following the principles of contingency planning. Governments retain significant room to manoeuvre and cope with extreme situations. Overall, water resources belong to the public domain and users are always subject to command and control measures and allocation criteria imposed by the government.

There are noticeable differences between these countries in terms of their use of water markets as a drought risk management tool:

- Canada, Spain and Australia allow voluntary water exchanges among users, whereas Australia has a clearer determination to make water markets liquid active and efficient.
- In the United Kingdom, abstraction trading is possible but not straightforward or quick. Short-term exchanges are generally not feasible under standard procedures due to the time required for approval.
- France is currently undertaking a reform to decentralise the management of water rights to farmer organisations, but there is at present no possibility to trade water rights either in the short-term nor in the long-term.

Australia's clear distinction between water use rights and allocations, both being tradable, ensures that water allocation is more market-driven and flexible. Water markets are pivotal during droughts. The Spanish water market is more heavily regulated and less active. But collective and public agencies can also react quickly, proactively and efficiently, like the French *agences de bassins* and Spanish river basin agencies, to reduce allocations or quotas. Comparison across countries in that regard suggest possible rooms for efficiency improvement based on the adoption of a more flexible, market-based approach to managing water scarcity and water shortages associated with droughts. A fair comparison, however, would deserve a proper cost-benefit analysis tailored to country-specific circumstances, which is beyond the scope of the present report.

Compensation and insurance policies against droughts is another area in which one can observe significant differences. Canada has risk management instruments that provide compensation for farmers' reduced margins, with respect to their individual records, irrespective of the cause, including reduced harvests caused by droughts (e.g. AgriInvest and AgriStability). As well, the AgriInsurance program offers multi-peril crop insurance that compensates for reduced yields caused by drought. Public authorities in France have been subsidising for a decade an insurance system (multiperil crop insurance) that indemnifies farmers for reduced yields caused by a range of climatic events including droughts. When available in certain countries, insurance provides risk-sharing and risk-transfer means for drought risks, but only for farmers relying on precipitation (rainfed). None provides compensation for water shortages. The observed differences cannot solely be explained by climatic conditions (drought being frequent in Australia, for instance). It is recognised that drought risks are difficult to insure due to a complex combination of market failures. Hence, countries may have different approaches depending on their institutional trajectories, as well as different objectives in terms of providing compensation to farmers. At least the comparison illustrates the fact that governments should have a careful approach in this area, based on sound cost-benefit analysis and a social welfare perspective. Recent assessment of the possibilities to develop crop insurance in Australia provides an interesting illustration of such an approach based on economic analysis.

In Canada, France, and to some extent Australia there are instruments available to farmers to smooth their revenues across time as an element of the risk management toolbox. However, the design, policy mix, and degree of public support to these time smoothing instruments vary a great deal across countries: *ex ante* subsidised precautionary saving tools for risk management purpose (Canada, France,); income tax smoothing schemes with or without subsidies (Australia, Canada, France); or *ex post* subsidised interest rates to refinance farms in circumstances of natural disaster (Australia, France). The overall policy coherence between time smoothing instruments and insurance and compensation instruments is, to some extent, difficult to characterise in all five countries.

Flood policies: Putting agricultural land on board can increase the cost-efficiency of flood risk mitigation

Floods events are characterised by their suddenness, but can also be predicted and anticipated. This feature sometimes tends to hinder the differentiation between crisis and recovery policies. Obtaining accurate mapping of flood risks is a prerequisite to design flood risk policies. In Australia, Canada, France, Spain and United Kingdom, different zoning instruments are put in place to delimit flood risk areas. In addition, information and monitoring tools offer information related to river levels, flow velocity and likelihood of flooding. Risk-layering is clearly developed in the five countries. Monitoring systems give information related to rainfall and river levels, which trigger the needed risk management actions.

Agricultural land use is a crucial element in the management of and protection from floods. Its role will be even more important in the coming decades when flood events could be more frequent and severe. The relative role of agricultural land in flood mitigation and floodplains varies a great deal across the five countries. In all, however, agricultural land areas are to varying degrees already included in the framework of general risk prevention policies against floods, with the underlying philosophy of an integrated flood risk management approach at the relevant scale, i.e. the water basin. It is difficult to disentangle specific elements related to agricultural land, in that regard. Some countries, however, such as France or United Kingdom, are increasingly considering the special role agriculture can play to mitigate flood risk in a cost-efficient way following a green infrastructure approach. Policy instruments can go beyond regulatory requirements and risk mapping, and include economic incentives through payments for ecosystem services, eventually bundled with other ecosystem services in view of maximising the social benefits from land use at the relevant spatial scale.

The insurance industry has been reluctant to offer coverage for flood risks in some countries. This is the case in Australia, where multi-peril crop insurance has been considered. In the United Kingdom, crop insurance for weather-related risks is limited. Farmers in Spain can purchase insurance against floods under certain regulatory conditions. Australia, France and Spain provide farmers compensation for uninsurable risks, i.e. risks that are not covered by private insurance markets. But Spain requires that those farmers receiving compensation for uninsured damage must have purchased basic insurance contracts. In France, insurable crop damages are not compensated. After flooding, *ad hoc* payments are used in Australia, Canada, France, Spain and United Kingdom to compensate farmers affected by a flood event.

Notes

1. Under the programme *Restoring the Balance* in the Murray-Darling, the Australian Government can acquire water entitlements from seller farmers in view to allow them to meet environmental needs.
2. The dedicated website www.nwrm.eu provides information about European Union policy on NWRM.
3. http://ec.europa.eu/environment/water/flood_risk/pdf/Better%20Environmental%20Options%20for%20Flood%20risk%20management%20ANNEXE.pdf.

References

Ansar, A., B. Flyvbjerg, A. Budzier, and D. Lunn (2014), "Should we build more large dams? The actual costs of hydropower megaproject development", *Energy Policy,* Vol. 69, pp. 43-56.

Babusiaux, C. (2000), *L'assurance récolte et la protection contre les risques en agriculture*, Ministère de l'Agriculture et de la Pêche, Ministère de l'Économie, des Finances et de l'Industrie, Paris. Available at: http://archives.agriculture.gouv.fr/publications/rapports/l-assurance-recolte-et-la-protection-contre-les-risques-en-agriculture/downloadFile/FichierAttache_1_f0/rapportbabuziaux-0.pdf

Brauman, K.A., S. Siebert, and J.A. Foley (2013), "Improvements in crop water productivity increase water sustainability and food security—a global analysis", *Environmental Research Letters*, N°8 024030.

Burek P., S. Mubareka, R. Rojas, A. de Roo, A. Bianchi, C. Baranzelli, C. Lavalle, I. Vandecasteele (2012), *Evaluation of the effectiveness of Natural Water Retention Measures*, JRC Scientific and Policy Report, European Union, Brussels.

Claassen, R., F. Carriazo, J.C. Cooper, D. Hellerstein and K. Udea (2011), "Grassland to Cropland Conversion in the Northern Plains: The Role of Crop Insurance, Commodity, and Disaster Programs", ERR-120, U.S. Department of Agriculture, Economic Research Service.

Combe, C. (2004), "Le risque d'inondation à l'amont de Lyon: héritages et réalités contemporaines", *Géocarrefour* Vol. 79, N°1, pp. 63-73.

Duflo, E. and R. Pande (2007), "Dams", *The Quarterly Journal of Economics*, Issue 122, No.°2, pp. 601-646.

Erdlenbruch, K., et al. (2009), "Risk-sharing policies in the context of the French Flood Prevention Action Programmes", *Journal of Environmental Management*, Volume 91, Issue 2, pp. 363-369.

Environment Agency (2007), R&D *Update review of the impact of land use and management on flooding*. Environment Agency, London.

European Commission (2013), "Green Infrastructures (GI) – Enhancing Europe's Natural Capital", COM(2013)249final, Brussels.

European Commission (2012), "A Blueprint to Safeguard Europe's Water Resources", COM(2012)673final, European commission, Brussels.

European Commission (2011), Towards Better Environmental Options for Flood Risks Management, European commission, Brussels, http://ec.europa.eu/environment/water/flood_risk/better_options.htm.

European Commission (2010), "Report on "dams and floods in Europe. Role of dams in flood mitigation", European Working Group on Dams and Floods, Luxembourg.

European Commission (2007a), "Drought management plan report: Including Agricultural, Drought Indicators and Climate Change Aspects", European Commission, Water Scarcity and

Drought Expert Network, Luxembourg, http://ec.europa.eu/environment/water/quantity/pdf/dmp_report.pdf.

European Commission (2007b), *Mediterranean water scarcity and drought report. Technical report on water scarcity and drought management in the Mediterranean and the Water Framework Directive*, Technical Report - 009 – 2007, Mediterranean Water Scarcity and Drought Working Group, www.emwis.net/topics/WaterScarcity/PDF/MedWSD_FINAL_Edition.

European Commission (2007c), *Communication from the Commission to the Council and the European Parliament, "Addressing the challenge of water scarcity and droughts in the European Union"*, Brussels, 18.07.07, COM(2007)414 final.

FAO (2008), "Coping with Water Scarcity: an Action Framework for Agriculture and Food Security", *FAO Water Reports* N°38, FAO Publishing, Rome.

Foudi S. (2013), *Flood risk management: what are the main drivers of prevention?*, Policy Briefing 02-2013, Basque Centre For Climate Change, Bilbao, www.researchgate.net/publication/271514790_Flood_risk_management_what_are_the_main_drivers_of_prevention.

Gómez, C.M. and Pérez-Blanco, C.D. (2014), "Simple Myths and Basic Maths About Greening Irrigation", *Water Resources Management*, Vol. 16 No, pp. 4035-4044.

Hoekstra A.Y., A.K.. Mekonnen, M.M., Chapagain, R.E. Mathews, B.D. Richter (2012), "Global Monthly Water Scarcity: Blue Water Footprints versus Blue Water Availability" *PLoS ONE,* Vol. 7(2).

Howitt, R.E., J. Medellin-Azuara, D. MacEwan, J.R. Lund, and D.A. Sumner (2014), *Economic Analysis of the 2014 Drought for California Agriculture*, Center for Watershed Sciences, University of California, Davis, California.

Jensen, C.R., J.E. Ørum, S.M. Pedersen, et al. (2014), "A Short Overview of Measures for Securing Water Resources", *Journal of Agronomy and Crop Science*, Vol. 200(5), pp. 333-343.

Kenyon W., G. Hill and P. Shannon (2008), "Scoping the role of agriculture in sustainable flood management", *Land Use Policy*, Vol. 25, pp. 351-360.

Kirby, M., R. Bark, J. Connor, M.E. Qureshi, and S. Keyworth (2014), "Sustainable irrigation: How did irrigated agriculture in Australia's Murray–Darling Basin adapt in the Millennium Drought?" *Agricultural Water Management*, in Press.

Kydland, F.E. and E.C. Prescott. (1977), "Rules rather than discretion: The inconsistency of optimal plans", *The Journal of Political Economy*, Vol. 85 pp. 473-491, www.jstor.org/stable/1830193.

Lankoski, J., et al. (2015), "Environmental Co-benefits and Stacking in Environmental Markets", *OECD Food, Agriculture and Fisheries Papers*, No. 72, OECD Publishing, Paris. http://dx.doi.org/10.1787/5js6g5khdvhj-en

Lefebvre M. and S. Thoyer (2012), *Risque sécheresse et gestion de l'eau agricole en France*, Laboratoire Montpelliérain d'Économie Théorique et Apliquée, Études et Synthèses, Montpellier.

Levido, L., D. Zaccaria, R. Maia, E. Vivas, M. Todorovic, and A. Scardigno (2014), "Improving water-efficient irrigation: Prospects and difficulties of innovative practices", *Agricultural Water Management*, Vol. 146, pp. 84-94.

Linnerooth-Bayer J, A. Dubel, J. Damurski, D. Schroeter, J. Sendzimir (2013), "Climate change mainstreaming in agriculture: Natural water retention measures for flood and drought risk management", *Policy Update* No. 7, FP7 RESPONSES Project - European Responses to Climate Change, Brussels.

Mallawaarachchi, T. and A. Foster (2009), "Dealing with irrigation drought: the role of water trading in adapting to water shortages in 2007-08 in the southern Murray-Darling Basin", Report prepared for the Australian Government Department of the Environment, Water, Heritage and the Arts (DEWHA), Australian Bureau of Agricultural and Resource Economics.

Morris, J., T. Hess and H. Posthumus (2010), "Agriculture's Role in Flood Adaptation and Mitigation: Policy Issues and Approaches", in OECD, *Sustainable Management of Water Resources in Agriculture*, OECD Publishing, Paris. http://dx.doi.org/10.1787/9789264083578-9-en.

Mortemousque, D. (2007), *Une nouvelle étape pour la diffusion de l'assurance récolte*, Ministère de l'Agriculture et de la Pêche, Paris.

O'Connell E., J. Ewen, G. O'Donnell and P. Quinn (2007), Is there a link between agricultural land-use management and flooding? *Hydrology & Earth System Sciences*, Vol. 11(1), pp. 96-107.

OECD (2015a), *Water Resources Allocation: Sharing Risks and Opportunities*, OECD Publishing, Paris. http://dx.doi.org/10.1787/9789264229631-en

OECD (2015b), *Drying Wells, Rising Stakes: Towards Sustainable Agricultural Groundwater Use*, OECD Studies on Water, OECD Publishing, Paris. http://dx.doi.org/10.1787/9789264238701-en.

OECD (2014), *Climate Change, Water and Agriculture: Towards Resilient Systems*, OECD Studies on Water, OECD Publishing, Paris, http://dx.doi.org/10.1787/9789264209138-en.

OECD (2013), *Environment at a Glance 2013: OECD Indicators*, OECD Publishing, Paris. http://dx.doi.org/10.1787/9789264185715-en.

OECD (2011), *Managing Risk in Agriculture: Policy Assessment and Design*, OECD Publishing, Paris. http://dx.doi.org/10.1787/9789264116146-en.

OECD (2010a), *Sustainable Management of Water Resources in Agriculture*, OECD Studies on Water, OECD Publishing, Paris. http://dx.doi.org/10.1787/9789264083578-en.

OECD (2010b), *Guidelines for Cost-effective Agri-environmental Policy Measures*, OECD Publishing, Paris, http://dx.doi.org/10.1787/9789264086845-en.

OECD (2009), *OECD Member Country Questionnaire Responses on Agricultural Water Resource Management*, OECD, Paris, www.oecd.org/agriculture/44763686.pdf.

OECD (2006), *Cost-Benefit Analysis and the Environment: Recent Developments*, OECD Publishing, Paris, http://dx.doi.org/10.1787/9789264010055-en.

Pattison I. and S.N. Lane (2011), "The link between land-use management and fluvial floodrisk: A chaotic conception?", *Progress in Physical Geography*, Vol. 36(1), pp. 72–92.

Pereira, L.S., I. Cordery and I. Iacovides (2012), "Improved indicators of water use performance and productivity for sustainable water conservation and saving", *Agricultural Water Management*, Vol. 108, pp. 39-51, http://econpapers.repec.org/article/eeeagiwat/v_3a108_3ay_3a2012_3ai_3ac_3ap_3a39-51.htm.

PREEMPT (2012), "Report on data and risk assessment methods used in reference river basins, policy requirements and potential for improvement", Deliverable B1, http://ec.europa.eu/echo/files/civil_protection/civil/prote/pdfdocs/fin_instr/projects2010/PREEMPT_final_report.pdf.

Reynaud, A. (2009), "Adaptation à court et long terme de l'agriculture au risque de sécheresse : une approche par couplage de modèles biophysiques et économique", *Revue d'Études en Agriculture et Environnement*, Vol. 90, No.°2, pp. 121-154.

Scheierling, S.M., D.O. Treguer, and J.F. Booker (2015), "Water productivity in agriculture: Looking for water in the agricultural productivity and efficiency literature," *Water Economics and Policy*. Forthcoming.

Schilling K.E., P.W. Gassman, C.L. Kling, et al. (2013), "The potential for agricultural land use change to reduce flood risk in a large watershed", *Hydrological Processes*, Vol. 28, pp. 3314–3325

Sidibé Y., J.-P. Terreaux, M. Tidball and A. Reynaud (2012), "Coping with drought with innovative pricing systems: the case of two irrigation water management companies in France", *Agricultural Economics*, Vol. 43, pp. 141-155.

Tirole, J. (1988), *The Theory of Industrial Organization*, MIT Press, Cambridge.

URSUS Consulting (2013), "Payment for Ecosystem Services (PES) Pilot on Flood Regulation in Hull", Report for DEFRA, www.google.fr/url?sa=t&rct=j&q=&esrc=s&source=web&cd=1&ved=0ahUKEwjAqobl4tvJAhVMtBoKHXnoC8UQFggcMAA&url=http%3A%2F%2Frandd.defra.gov.uk%2FDocument.aspx%3FDocument%3D11136_FinalreportHullPESPilot130513.pdf&usg=AFQjCNFMCsrXZavhwJ43r31pRcAMueHTiQ&bvm=bv.109910813,d.ZWU.

UK Environment Agency (2010), *Working with Natural Processes to Manage Flood and Coastal Erosion Risk*, London, http://webarchive.nationalarchives.gov.uk/20140328084622/http:/cdn.environment-agency.gov.uk/geho0310bsfi-e-e.pdf.

World Bank (2010), "Assessment of Innovative Approaches for Flood Risk Management and Financing in Agriculture", *Agriculture and Rural Development Discussion Paper N°46*, Washington, DC.

World Meteorological Organization (2009), *Integrated Flood Management Concept Paper*, WMO.

Annex 3.A1

Drought and flood prevention policies in OECD countries

Table 3.A1.1. Prevention policies against droughts in OECD countries, 2009

Policies	Instruments	Countries
Regulatory instruments		
Planning	National Drought Strategy or Water Efficiency Programme	Australia, Canada, Portugal, United Kingdom
	Water users associations and farm water plans	Netherlands, Portugal, Turkey
Minimum environmental flows	Benchmarking in water channels	Australia, Austria, Denmark, Finland, France, Germany, Greece, Italy, Japan, Korea, New Zealand, Poland, Portugal, Spain, Switzerland, United Kingdom, United States
Economic incentive measures		
Water retention and storage	Off-farm irrigation infrastructure	Australia, France, Greece, Hungary, Italy, Korea, Poland, Portugal, Slovak Republic, Spain, Turkey, United States
	Upgrades of on-farm irrigation/water storage	Australia, Canada, Greece, Italy, Poland, Portugal, Switzerland, Turkey
	Water recycling in fields or paddy rice fields	Japan, Korea
	Desalinisation	Spain
	Recycled treated sewage	Portugal, Spain
Soil moisture retention	Agri-environmental practices (e.g. conservation tillage, restoration of terraces, change cropping systems)	Canada, Czech Republic, France, Greece, Hungary, Poland, United States
	Wetland restoration and conservation	All OECD countries except Turkey
Other policy instruments		
Farm advisory, famer technical guidance and education	Educational and advisory programmes	Australia, Austria, Canada, Hungary, New Zealand, Portugal, United States
Information	Research	Australia, Austria, Belgium, Portugal
	Water scarcity indicators	Portugal, Spain

Source: OECD Member Country Questionnaire Responses on Agricultural Water Resource Management (2009).

Table 3.A1.2. Prevention policies against floods in OECD countries, 2009

Policies	Instruments	Countries
Regulatory instruments		
Planning	Restrictions in high risk zones	Austria, Finland, France, Ireland, Poland, Portugal, Spain
	Considerations of changes in land use management on runoff	Belgium, Netherlands, Portugal
Economic incentive measures		
Water retention and storage	Upland runoff retention	Belgium, Finland, France, Poland, Spain
	Lowland water storage: polders, washlands	Austria, Belgium, Czech Republic, France, Hungary, Netherlands, Portugal, United Kingdom, United States
	Best management practices; e.g. agri-environmental schemes	Czech Republic, France, Greece, Hungary, Netherlands, Portugal, United Kingdom
	Compensation for water retention	Austria, Belgium, Hungary, Netherlands
	Wetlands	Belgium, Finland, France, Hungary, Poland, Sweden, United States
	Erosion control	Belgium, Czech Republic, France, United States
	Afforestation	Poland, Portugal
Other policy instruments		
Flood defence agricultural areas	Increase flood defence	Belgium, Czech Republic, Greece, Hungary, Poland, Portugal, Spain
	Withdrawal of flood defence	Austria, Finland, Ireland, Portugal, United Kingdom
	Land drainage	Poland, Portugal, Sweden
	River restoration	Austria, Canada, France, Netherlands, Portugal, Switzerland, United Kingdom
Land use	Lowland water storage: paddy fields	Japan, Korea
Information	Flood risk mapping	Austria, Belgium, Japan, Portugal, Spain
	Flooding warning systems	Belgium, Spain, Portugal, United Kingdom

Source: OECD Member Country Questionnaire Responses on Agricultural Water Resource Management (2009).

Annex 3.A2

Synoptic tables of the main characteristics of policy approaches to droughts and floods in agriculture: Australia, Canada, France, Spain and the United Kingdom

Table 3.A2.1. Schematic representation of water policies in Australia

	Water policies in Australia	
	Drought policies	**Flood policies**
Long-run water risk mitigation	• Water entitlements with different reliability levels, and water allocations • Decentralised allocation systems in some basins (storage rights, capacity sharing system) • Different water pricing mechanisms among regions • Structural measures; State water infrastructure measures	• Land-use permits
Prevention	• Information tools • Farm Management Deposits • Farm business training • Promotion of water use efficiency and conservation • Farm Management Deposit Scheme (for all risks) • State pest management initiatives	• Information tools • Flood warning services • Land use controls • National Partnership Agreement on Natural Disaster Resilience • Regional programs to reduce flood-related risks
Short-run water risk management	• Drought maps highlight areas suffering from serious or severe rainfall deficiency • Water entitlements reliability levels • Water markets (permanent or spot) • Water saving measures • Rural Financial Counselling Service • Broader rural assistance policies (Farm Household Allowance)	• Sustainable Farm Families Program • Clean-up and restoration grants
Compensation	• Farm Household Allowance • Drought Recovery Concessional Loans Scheme • Social and community support measures	• National Disaster Relief and Recovery Arrangements (policy and financial assistance)
Risk thresholds	• Based on two levels of severity (or return periods)	• Not clearly defined for risk-layering
Risk-layering	Range of assistance measures available to drought-affected farmers depending on particular circumstances	Predictions of the expected height of a river and the time that this height is expected to be reached

Table 3.A2.2. Schematic representation of water policies in Canada

	Water policies in Canada*	
	Drought policies	**Flood policies**
Long-run water risk mitigation	• Water licenses/permits (each province has its own system) • Water pricing • Structural measures	
Prevention	• Information tools (Drought Watch, National Agroclimate Information Service, agriculture drought severity maps) • Promotion of practices to reduce drought vulnerability and improve management • Technical and financial assistance to switch to more secure water supplies	• Monitoring • Flood preparedness activities • Flood Protection Program (funding for mitigation strategies)
Short-run water risk management	• Water markets • Water rationing	Range of provincial response can vary dramatically depending on the severity of the event
Compensation	• Disaster Financial Assistance Arrangements for provincial and territorial governments • AgriInsurance plans developed and delivered by each province to meet the needs of local producers • AgriRecovery • AgriStability	• Disaster Financial Assistance Arrangements for provincial and territorial governments • AgriInsurance plans developed and delivered by each province to meet the needs of local producers • AgriRecovery • *Ad hoc* payments • AgriStability
Risk thresholds	• Not clearly established	• Not clearly defined for risk-layering
Risk-layering	Real-time reports of lake and reservoir levels, stream flows, snowpack accumulations, water-supply volume forecasts, dugout water levels (for the Prairies), and precipitation anomalies	Three levels of advisory based on river levels and flooding of adjacent areas: high streamflow advisory, flood watch and flood warning.

* Due to the great diversity in water planning in Canada, as each province is responsible for this, some of the information on drought policies in this table is related to the province of Alberta in particular, and for British Columbia in case of flood policies.

Table 3.A2.3. Schematic representation of water policies in France

	Water policies in France	
	Drought policies	**Flood policies**
Long-run water risk mitigation	• Water quotas based on the crop portfolio in hectares • Water pricing • Structural measures • Management performed by collective units of water management • Criterions for abstraction permits	• Enforcement of land use codes
Prevention	• Information tools • Monitoring • Support to water saving technologies	• Information tools • Monitoring • Risk Prevention Plans • PAPI programs (finance investments to transfer flooding flows from high to lower vulnerability areas) • Mapping of flood hazard areas
Short-run water risk management	Drought decrees to provisionally limit or even prohibit abstractions	After an alert declaration, several preparedness and response plans are activated, together with actions undertaken by different responsible bodies
Compensation	• *Fonds National de Garantie des Calamités Agricoles* (National Guarantee Fund for Agricultural Disasters) • Multi-peril crop insurance	• Payments from the *Fonds National de Garantie des Calamités Agricoles* (National Guarantee Fund for Agricultural Disasters) • Multi-peril crop insurance
Risk thresholds	• They are defined for the purpose of risk-layering	• Defined for risk-layering and cartographies of flood areas
Risk-layering	Different alert levels based on river flows and groundwater levels Thresholds leading to restrictive measures are defined locally by the prefects. This risk-layering system facilitates the reaction in crisis situations, and enables transparency and consultation between the different users of the same basin	Rainfall (mm), depending on the regional climatology

Table 3.A2.4. Schematic representation of water policies in Spain

	Water policies in Spain	
	Drought policies	**Flood policies**
Long-run water risk mitigation	• Water use rights • Water pricing • Structural measures	• Enforcement of land use codes • Permanent Commission for Climatic and Environmental Hazards
Prevention	• National Drought *Observatory* • State Meteorological Services • Programs to improve irrigation systems and enhance water savings • Integration of climate change projections in River Basin Plans • Drought *Management* Plans, approved	• State Meteorological Services • National System of Flood Zone Mapping (SNCZI) • Environmental measures • Flood risks and flood hazard map • Urban and land use planning • Regular Risk assessments
Short-run water risk management	• Drought *decrees* • Water use restrictions • Water *markets* • '*Drought* wells' • National Drought Committee is set up to coordinates all efforts	• Civil protection • Operations and management of flows and releases in a real-time basis
Compensation	• Crop *insurance* • *Compensation* Funds • Special funding mechanisms approved by the *national* and autonomous government in order to offer economic compensation to farmers affected by droughts and fast-track funding for emergency infrastructures and other supply measures	• Crop insurance • The National government can trigger different urgent and exceptional complementary measures to help in the recovering from the damages caused by severe flooding.
Risk thresholds	• They are defined for the purpose of risk-layering	• Defined for risk-layering and cartographies of flood areas
Risk-layering	Clearly defined by the Drought *Management* Plans and in drought (yields) insurance instruments The SAIH supports the formal declaration of alert *situations* (floods and droughts) by River Basin Agencies	• In each basin, the SAIH collects the information related to river flows and define internal alert levels

Table 3.A2.5. Schematic representation of water policies in the United Kingdom

	Water policies in the United Kingdom	
	Drought policies	**Flood policies**
Long-run water risk mitigation	• Water licenses • Water pricing • Structural measures	• Enforcement of land use codes
Prevention	• Monitoring • Environment Agency's regional water situation report.	• Monitoring • Flood maps • Environment Stewardship schemes • Flood and Coastal Erosion Resilience Partnership Funding • Environment Agency's regional water situation report. • 'Making Space for Water' (funding for rural land use solutions such as the creation of wetlands and washlands, and managed realignment of coasts and rivers) • Flood Schemes
Short-run water risk management	• Drought permits, ordinary drought orders and emergency drought orders • Spray irrigation restrictions • Water markets	During the flooding, the involved parties in flood risk management work together to assist the affected people and coordinate the emergency response.
Compensation	• Agri-environment schemes • Measures to help farmers to fill their water reservoirs	• Disaster relief funds • Emergency funds • Practical and emotional support • Addington Fund to help farmers get vital supplies • Farming Recovery Fund
Risk thresholds	• Defined for droughts	• Not clearly defined
Risk-layering	Regional plans contain the actions that should be taken at different drought stages and detail the indicators that will determine these actions	To assess the level of risk a large number of weather, catchment and coastal factors are taken into account. These are then presented on a coloured risk matrix depending on the likelihood and impact

ORGANISATION FOR ECONOMIC CO-OPERATION AND DEVELOPMENT

The OECD is a unique forum where governments work together to address the economic, social and environmental challenges of globalisation. The OECD is also at the forefront of efforts to understand and to help governments respond to new developments and concerns, such as corporate governance, the information economy and the challenges of an ageing population. The Organisation provides a setting where governments can compare policy experiences, seek answers to common problems, identify good practice and work to co-ordinate domestic and international policies.

The OECD member countries are: Australia, Austria, Belgium, Canada, Chile, the Czech Republic, Denmark, Estonia, Finland, France, Germany, Greece, Hungary, Iceland, Ireland, Israel, Italy, Japan, Korea, Luxembourg, Mexico, the Netherlands, New Zealand, Norway, Poland, Portugal, the Slovak Republic, Slovenia, Spain, Sweden, Switzerland, Turkey, the United Kingdom and the United States. The European Union takes part in the work of the OECD.

OECD Publishing disseminates widely the results of the Organisation's statistics gathering and research on economic, social and environmental issues, as well as the conventions, guidelines and standards agreed by its members.

OECD PUBLISHING, 2, rue André-Pascal, 75775 PARIS CEDEX 16
(51 2015 18 1 P) ISBN 978-92-64-24673-7 – 2016

www.ingramcontent.com/pod-product-compliance
Lightning Source LLC
LaVergne TN
LVHW081421110826
845149LV00010B/1831

* 9 7 8 9 2 6 4 2 4 6 7 3 7 *